Coexisting with Large Carnivores

Coexisting with Large Carnivores

Lessons from Greater Yellowstone

Edited by

Tim W. Clark

Murray B. Rutherford

Denise Casey

ISLANDPRESS

Washington • Covelo • London

Library of Congress Cataloging-in-Publication data.

Coexisting with large carnivores : lessons from Greater Yellowstone / Edited by Tim W. Clark, Murray B. Rutherford, Denise Casey.
p. cm.
Includes bibliographical references and index.
ISBN 1-59726-005-3 (pbk. : alk. paper) — ISBN 1-59726-004-5 (hardback : alk. paper)
1. Human-animal relationships—Yellowstone National Park Region. 2. Carnivora—Behavior—Yellowstone National Park Region. 3. Wildlife management—Yellowstone National Park Region. I. Clark, Tom W. II. Rutherford, Murray B. III. Casey, Denise.
QL85.C64 2005
639.97'97—dc22

2005006225

British Cataloguing-in-Publication data available.

♾ Printed on recycled, acid-free paper

Design by Paul Hotvedt, Blue Heron Typesetters
Photos on Frontispiece and Part One, Two, and Three opening pages are courtesy of Tom Mangelsen

Manufactured in the United States of America
10 9 8 7 6 5 4 3 2 1

Contents

Preface

This volume will be of interest to anyone involved with or concerned about the management and conservation of large carnivores in Greater Yellowstone, North America, or elsewhere in the world. Learning to coexist with large carnivores—that is, conserving their populations and ecosystems over the long term, while at the same time allowing humans and human communities to thrive—is not an easy task. Like many other resource management problems, it is fraught with intense conflict, historical baggage, and complexity on multiple levels. However, we have an opportunity and an obligation to learn the skills needed for coexistence *now,* at what may be the eleventh hour for many carnivores and their ecosystems. We hope that this volume will encourage managers, researchers, government officials, ranchers, and anyone else who is affected by problems associated with large carnivores to redouble their efforts and put in place workable, democratic means to resolve differences and find common ground. In the end, we all want outcomes that are reasonable, practical, and morally justified.

There are many places in the world where the drama and calculus of living with large carnivores are being worked out in people's daily lives. This book focuses on one significant such "laboratory" for learning about coexistence—the portion of western Wyoming that lies adjacent to and south of Yellowstone National Park, including the southern part of what is commonly recognized as the Greater Yellowstone Ecosystem (see figure 1.1 in chapter 1). Here, the problems of living with large carnivores are currently being faced on the ground on a day-to-day basis. Yellowstone National Park still supports populations of grizzly bears and mountain lions, and wolves were reintroduced in 1995. Recovery programs for grizzlies and wolves have been successful enough that many of these animals are now moving out of protected areas south into territory that is dominated by human uses. Predictably, there have been conflicts. Cattle and sheep have been attacked, hunters have shot bears when they perceived that their own safety was threatened, and "problem" bears and wolves have been killed by wildlife managers. Mountain lions have also become a focal point for public concern, and traditional approaches to the management and hunting of lions are now extremely controversial. These interactions with large carnivores are being played out in an emotionally charged and highly symbolic political maelstrom, as the culture and institutions of the Old West attempt to deal with the modern reality of economic transition and an influx of wealthy urbanites.

The western Wyoming story is fascinating for many reasons. It is a story of struggle, in which people must learn to live with one another, resolve their differences, and harmonize their practices with the requirements of nature. This struggle is a microcosm of the larger search for coexistence and sustainability in the United States, across North America, and throughout the world. We can learn much from experiences in this region that can be applied to other sites and efforts to increase the likelihood that people can learn to live with large carnivores.

For many people in the counties that abut Yellowstone National Park and the surrounding national forests, large carnivores are simply bad and should be put out of the way of human progress. Some advocate restricting these animals by relegating them to remote areas. Several Wyoming counties have actually declared the grizzly bear a "socially and economically unacceptable species within our counties."[1] The wolf and, to a lesser extent, the mountain lion, are also viewed with hostility. Many local citizens and officials alike in this region feel that the federal government, through the Endangered Species Act and other laws, and the state of Wyoming, through its wildlife management programs, have "forced these large predators, and the resulting regulations, upon our counties."[2]

At the same time, there are people on the other side of the debate who have organized to achieve greater protection for large carnivores. For them, carnivores have a rightful place and should be allowed to roam and prosper. Clearly, these animals mean or symbolize different things to different groups of people. Getting these opposing camps together in a room to discuss carnivore management can be incendiary. People resort to shouting bitter accusations at their neighbors. "Facts" are used selectively to support one side of the debate or the other. Dramatic incidents of bears or wolves eating livestock loom large. When these angry exchanges take place at public meetings, agency officials are often at a loss about what to do or how to turn the situation from ranting and raving into a constructive forum in which citizens seek common ground together. When all is said and done, no one seems to be satisfied with the current situation or its outcomes. Neither anti- nor pro-carnivore people are likely to get what they want under these conditions.

But what can be done? Where is the balance? How can we find ways to coexist with large carnivores? How can we manage the conflict among ourselves and find workable outcomes? The fundamental question is, "Can people work together to find sustainable solutions?" These questions must be addressed in practical terms if we are to find satisfactory answers for human and wildlife communities. The people involved in large carnivore management must decide ultimately if they want to continue shouting past one another, or sit down instead and deal realistically with their very real problems.

This book helps to untangle some of these highly charged issues. We recommend steps to break the current cycle of corrosive conflict and reverse the erosion of social capital. Specifically, we suggest strategies to resolve actual, on-the-ground conflicts with carnivores more effectively, change what these animals symbolize or mean to people, and improve the institutional system of wildlife management to operate in a more timely, fair, and effective manner. Although much has been written elsewhere about the ecology of large carnivores and the problems of managing populations of these species to ensure recovery within protected areas, little has been written that adequately addresses the sociopolitical problems of coexisting with large carnivores outside of protected areas. This book is intended to fill that gap.

Many people are looking for practical ways to improve policies and practices for natural resource management. In our view, the only way through the current morass is a mutually respectful, collaborative, problem-solving approach. We hope that the insights and ideas that come from our look at this part of Greater Yellowstone will help to reinvigorate people's commitment to working together to find democratic solutions to environmental problems.

About This Volume

The book is divided into three parts. Part I begins with an introduction to large carnivore management as a social process (chapter 1), examining the ongoing struggle and its significance to the future health of the region and its human and animal inhabitants and identifying fundamental problems that must be overcome to achieve the goal of coexistence. Following this "problem orientation," chapter 2 maps the context in which these problems exist, emphasizing the practices and beliefs that have brought large carnivores near to extinction and those that could open up avenues for improvements. Part II consists of three chapters, each of which is a case study of one of the species at issue (mountain lions, grizzly bears, and wolves). Each case study begins with a review of the natural history, population dynamics, and management history of the species. Then the authors discuss the ways in which management has become increasingly politicized and tied to broader sociopolitical issues. Finally, each chapter derives lessons from the case at hand and evaluates options for improving conservation.

Part III focuses in greater detail on two key dimensions of the carnivore management problem. Chapter 6 examines community-based, participatory processes and what can be done to upgrade citizen participation and democracy-in-action in the coexistence debate. Chapter 7 takes an institutional view, analyzing and evaluating the institutional system of wildlife management and making recommendations for improvement. Finally, the concluding chapter summarizes the lessons to be learned from experiences with large carnivores in this region and the relevance of these lessons for other settings. Throughout the book the authors emphasize effective, joint problem solving and helping people see past their individual special interests to find the common interest.

Origin of Our Work

This book began with a joint project by Greg McLaughlin, Karen Murray, Lyn Munno, Dylan Taylor, and Jason Wilmot in fall 2000 for a graduate seminar with Tim Clark at the Yale University School of Forestry and Environmental Studies. They looked at the interactions among wildlife, human communities, and the institutional arrangements or "policy systems" that determine large carnivore management. Their initial efforts—library research, telephone interviews with wildlife professionals, newspaper articles, and in-depth analyses—used an interdisciplinary approach, drawing on insights from psychology, sociology, political science, and organizational and policy literature. This was followed up with face-to-face conversations with people on all sides of the conflict, those who participate in or influence the decision-making process and those who are affected by it.

From discussions about grizzlies, wolves, and mountain lions with more than 40 state and federal agents, ranchers, hunting outfitters, scientists, and conservationists, these researchers gathered detailed and accurate information on carnivore management and areas of conflict. They also looked for consistencies and inconsistencies between key players' verbal accounts and the publicized and observable accounts of their actions. Throughout the project, they kept in mind the goal of collaborative management and looked for opportunities to help. The data they gathered and the insights they gained from their research and discussions inform all of the analyses in this volume.

Acknowledgments

This project could not have been completed without the generous time, efforts, and funding of many individuals and organizations. We would like to thank the following individuals for taking time to speak and work with us: Dennis Almquist, Linda Baker, Kim Barber, Levi Broyles, Mark Bruscino, Franz Camenzind, Len and Anne Carlman, William Cramer, Lloyd Dorsey, Duke Early, Betty Fear, Barb Franklin, Dave Gaillard, Mark Gocke, Rachel Grey, Matt Hall, Kniffy Hamilton, Rick Hartley, Jennifer Hayward, Kathe Henry, Mark Hischberger, Bernie Holz, Robert Hoskins, Dan Ingalls, Mike Jimenez, Gordon Johnston, Les Jones, Maury Jones, Timm Kaminsky, Joette Katzer, Louise Lasley, Pam Lichtman, Eric Lindquist, Mary Maj, Tom Mangelsen, Susan Marsh, Steve and Sydney Martin, Dave Mattson, Brad Mead, Dave Moody, the Norm Pape family, Steve Primm, Barry Reiswig, Brian Remlinger, Bart Robinson, Jon and Debbie Robinett, Alan Rosenbaum, Jonathan Schechter, Carl Schneebeck, Terry Schram, Michael and Claudia Schrotz, Chris Servheen, Jamie Shane, Sandy Shuptrine, Tom Skeele, Wes Smith, Albert Sommers, Gary Tabor, Tory and Meredith Taylor, Tom and Judith Wiancko, Louisa Willcox, Seth Wilson, and Bruce and Mary Wolford.

We would like to extend a special thank you to Tom Mangelsen and Images of Nature for the use of Tom's superb photos throughout the book.

We are also grateful for the cooperation of the following organizations: Bridger-Teton National Forest, Shoshone National Forest, Diamond G Ranch, Wyoming Game and Fish Department, U.S. Fish and Wildlife Service, Upper Green River Cattle Association, Sublette County, Teton County, Northern Rockies Conservation Cooperative, Predator Conservation Alliance, National Elk Refuge, Jackson Hole Conservation Alliance, Jackson Hole Outfitters, Wyoming Wildlife Federation, Hornocker Wildlife Institute, Images of Nature, Grand Teton National Park, the Wyoming Outfitters and Guides Association, U.S. Department of Agriculture Natural Resources Conservation Service, and Yellowstone to Yukon Conservation Initiative.

This project relied on the generous support of the Northern Rockies Conservation Cooperative in Jackson, Wyoming; Yale University School of Forestry and Environmental Studies Class of 1980 Fund; and the Carpenter/Mellon/Sperry Fund. Other funds came from Catherine Patrick, Doug Smith, Bart Robinson, Gilman Ordway, the Wiancko Family Fund, Nancy Kittle, Steve and Amy Unfried, Yvon and Malinda Chouinard and the Wilburforce Foundation.

Several people critically reviewed the manuscript or portions thereof. Some reviews were anonymous and others were made by well known authorities in the field. We thank you all.

The Editors
Tim W. Clark
Murray B. Rutherford
Denise Casey
November 30, 2004

References

1. C. Urbigkit, 2002, "The coup counties: Western Wyoming commissioners outlaw grizzlies and wolves," *Range,* Fall, 30–31.

2. Urbigkit, "The coup counties: Western Wyoming commissioners outlaw grizzlies and wolves."

Part One:

Context

Chapter **1**

Coexisting with Large Carnivores: Orienting to the Problems

Tim W. Clark and Murray B. Rutherford

Large carnivores mean vastly different things to different people, and these meanings are often associated with intense feelings.[1] A sampling of quotes from articles and newspapers published in the American West makes this plain. For some people, large carnivores are an outlet for strong resentment about the course of recent history in the West. For example, Cat Urbigkit, a sheepherder and coowner of the *Sublette County Examiner* (a small newspaper published in Pinedale, Wyoming), reported in 2002 that county commissioners in several Wyoming counties had "outlawed" grizzly bears and wolves. "Fed up with mandates from the federal government," they took action, "adopting resolutions prohibiting the presence, introduction or reintroduction of grizzly bears and wolves within the boundaries of their counties." They "drew a line in the sand," by proclaiming that they would no longer tolerate these kinds of actions from the federal government. As Todd Wilkinson, a well-known writer on natural resource issues in the West, observed, some people just "hate wolves. They hate grizzlies. They hate government (except federal subsidies). They hate public education. They hate any law which constrains their 'personal liberty.'" They spin elaborate, sometimes slanderous, yarns about conservationists plotting to "'lock Americans out of public lands'; allegedly scheming to lure the U.S. into a 'one-world government,' headquartered by the United Nations; and finally [driving] all rural people off the land." Wilkinson suggested that "hatred of wolves could be a symptom"—we might also call it a symbol—"of something else: fear of losing control over things in our lives, which inherently are beyond our control."[2]

In dramatic contrast, the return of large carnivores has been a welcome event for other people, evidence that the region's ecology is "healing" and returning to the "way it should be." According to noted author Tom McNamee, putting wolves back into nature "may have saved the ecosystem from ruin." Similarly, Robert Ferris of Defenders of Wildlife observed, "Restoration of these animals represents a major step in correcting earlier errors in public policy and in repairing ecological imbalances." And Greg Hanscom in *High Country News* said, "Restore the top predator and you restore the entire ecosystem."[3]

These contrasting quotes make two things very clear. First, carnivores symbolize many other issues, and second, the differences among these "meanings" pose a serious practical management challenge.[4] Finding common ground is indeed an uphill task.

Nowhere is the conflict over meaning and management sharper than in the area of the West on which this book focuses—western Wyoming south of Yellowstone National Park, including the southern part of what is commonly recognized as the Greater Yellowstone Ecosystem. Greater Yellowstone is one of the largest and most important systems of protected areas in the United States and one of the more significant regions for conservation in the world. As figure 1.1 shows, the area covered by this book encompasses about 22,000 square miles and includes all of Bridger-Teton National Forest, other federal and state lands, and private lands. The overall boundaries are not rigidly defined, and large carnivores range widely in and out, but in recent years human-carnivore conflicts seem to be clustered in this area (see figure 1.1). This is the stage on which a public policy play is currently being acted out, and it can serve as a field laboratory for us to learn how to secure a future for large carnivores in a dynamic human context.

The difficulty of coexisting with large carnivores is less about the carnivores than it is about us and our views. The basic problem is how we go about interacting with one another over troubling public issues and collectively deciding how we want to live. We can manage large carnivores. However, it is much more challenging to manage ourselves in cooperative ways that will give large carnivores more room than they presently have.

Grizzlies, wolves, and mountain lions were eliminated or reduced to very low numbers in most of Greater Yellowstone decades ago. But

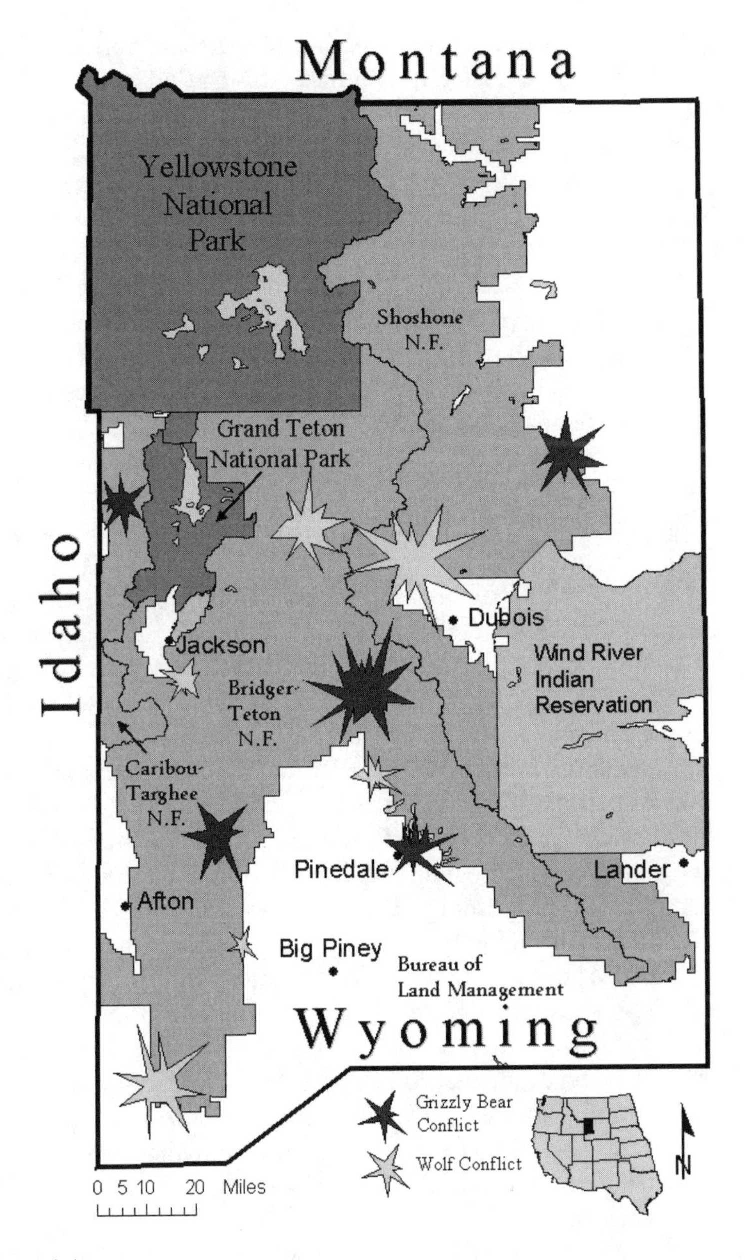

Figure 1.1
Map of the southern part of the Greater Yellowstone Ecosystem, showing Yellowstone National Park and cities to the south and selected sites of human-large carnivore conflict.

in recent years many Westerners, along with others nationwide, have called for the restoration of these animals. This has led to more active conservation, restoration, and reintroduction programs. Since the early 1990s, grizzly bears and wolves have moved south from Yellowstone country into western Wyoming, and the number of mountain lions may also have increased. This has led to conflicts among people about what to do with their new and sometimes unwanted neighbors, which occasionally eat sheep and cattle in addition to deer and elk. The intense feelings that people have about carnivores, the return of these animals to areas from which they have long been absent, and the new conflicts occurring among people and between carnivores and people have combined to make management of large carnivores a complex and messy political problem. It can also be highly personal and costly for some of the participants. Managers, ranchers, and environmentalists have occasionally been vilified or glorified in the media. They too have become political symbols.

So what are the real issues behind the symbolism? Can anything be done to change these interactions so that people and wildlife can live in sustainable coexistence with one another?[5] Grizzly researcher Steve Primm, with the Northern Rockies Conservation Cooperative, says the answer is decidedly "yes!" Steve is working to develop a community-based grizzly management process near Ennis, Montana, and is building food storage poles in backcountry campsites south of Yellowstone. He believes that we must deal with the real bears and the real problems they sometimes cause, while recognizing at the same time that these bears are highly symbolic (Primm elaborates on these views in chapters 4 and 6). Wildlife biologist Timm Kaminski also says "yes!" Now with the Mountain Livestock Cooperative, Timm is working to show people who want to protect bears, wolves, and mountain lions, and those who earn a living from the land, that people can and do solve difficult carnivore management problems by learning from each other. These two experienced field workers are among a growing number of individuals who believe that carnivore management can be much more effective in promoting and achieving coexistence between people and carnivores.

In this chapter and throughout this book we argue that to achieve coexistence with large carnivores we must think and act in ways that were unthinkable a few, short decades ago. We must minimize local, on-the-ground conflicts between people and predators, while finding ways to change what carnivores mean and symbolize. We must be adaptive and use "practice-based learning" to build on our past successes—drawing on experiences in actual situations to learn what works (and what does not work) to solve or minimize problems.

Carnivore Management as a Social Process

Managing large carnivores is a complex, dynamic, ongoing, social process. It directly reflects the feelings, beliefs, and values of the many people who participate in one way or another. Understanding this complex social process is a vital first step to envisioning how we can change things for the better.

At the Center: People and Their Perspectives

Although we often focus our attention on grizzlies, wolves, and mountain lions, we should never forget that people are involved. Many people think that carnivore management is a fairly cut-and-dried activity, carried out by technical experts (scientists and managers working for government) who are objective and neutral and who operate with the public interest in mind. In fact, carnivore management is an ongoing process in which many people—managers, scientists, ranchers, environmentalists, and those with other interests—make decisions about what we all value (although we don't all value the same things). Carnivore management is a political process that is only partially scientific. It is a "transscience" issue that involves science, but goes well beyond what science can offer. Symbols and symbolic victories are at least as important as real successes. Biases figure prominently, even on the part of experts.

To some people, carnivores should simply be destroyed. They argue that our frontier forebears virtually eliminated large carnivores half a century or more ago because they were so damaging to ranching and that, even though the populations of these animals are now much smaller, they continue to threaten livelihoods and pose unfair costs. These people claim that predators stand in the way of progress and should be eliminated, much reduced, or restricted to distant regions. To other people, though, these same animals symbolize "free nature" and a "healthy environment." They make the counterclaim that carnivores should be left alone to live in the "wilderness" with other wildlife "as they were meant to." They see carnivores as beneficial and say that local people should adjust to them, even if it means going out of business or ending generations of family tradition. These claims and counterclaims in the discourse about carnivore management show that strikingly different perspectives are at play, which are being symbolized in words, advocacy, and agency politics and programs. Consequently, conflict is typically at center stage when communities try to decide how to live with large carnivores.

Regardless of which side of the issue people are on, there is a tendency to label those on the other side as misguided, wrongheaded, ignorant, in need of education, or even malevolent or untrustworthy. Humans have a predisposition to stress group identity and exclusivity of membership and to use labels such as these to divide the world into "us vs. them." Terms such as "rancher" and "hunter" are examples of group identity labels that take on added meaning when contrasted with other labels such as "environmentalist" and "conservationist." Similarly, state agents may identify themselves in opposition to federal agents. The notion of "we and they" is the central theme that holds groups and societies together by creating individual and group meaning. Our core identities are formed around such groups, regardless of whether we tend to be parochial or cosmopolitan in our worldviews. This dynamic is clearly evident in large carnivore management.[6]

To be more successful in carnivore management, we must work with this dynamic of identity formation and be sensitive to the needs and wants of all the people and groups involved. We need to learn what is most important to people, how we can balance one group's demands with those of other groups, and when and where to apply leverage to get people to compromise, work together, and set their sights on common goals. To do this, we need to strengthen the institutions associated with carnivore management and build in the capacity to learn from on-the-ground experiences with people and animals.

The Institutional System of Wildlife Management: Meeting High Standards, Serving Common Interests

We often think of wildlife management as doing something good for animals or their habitats. But as the foregoing examination of human social process has made clear, actually we are managing not the animals, but ourselves. For example, it is people's behavior that we target when we decide not to kill carnivores, or log a forest, or run sheep without herders, or hike or hunt in grizzly country, or leave wilderness areas littered with human food. To understand carnivore management, then, we need to understand ourselves and how we make decisions through the institutions of wildlife management. As we discuss in more depth in chapter 7, an institution is a "well established and structured pattern of behavior or of relationships that is accepted as a fundamental part of a culture."[7] Among other things, institutions embody and prescribe the norms and rules for our decision making and actions. The many institutions associated with managing wildlife constitute a system that we call the "institutional system of wildlife management." For better or worse, it is this institutional system that we must work with to find a balance between people and predators. If this system fails, we must find ways to restructure it to serve people and nature better and to ensure a healthy future for ourselves.

The wildlife management process requires us to ask and answer many questions of ourselves. For example, what should be our goal—coexistence or elimination of carnivores? If we decide on coexistence, what do we mean by that, given the context in which we are operating? Should we limit our own actions that harm carnivores? If so, which ones, when, where, and how? How can we inspire adequate, constructive debate about these and other matters among all the people that matter? Should we work to improve our understanding of these species? What kind of scientific information—both biological about the animals, and social about ourselves—should we gather? How do we integrate this information so that we understand the situation and the problems realistically? What is the best way to learn about, to frame, or to define any problems? How can we be sure that we have fully explored all the options to fix the problems? What management practices should we carry out to ensure that we overcome the problems we have identified? How can we fairly distribute the burden of living with carnivores among all of the people and interests involved? What kinds of rules, plans, and actions are needed to guide and coordinate our work? How can we ensure that the community's decisions will be implemented promptly, fairly, and effectively? In what ways can we best monitor our actions and the responses of both people and carnivores to our decisions and their implementation? How do we reveal our own assumptions in all of this to ourselves, so we can take these assumptions into account in the entire decision-making and -implementing process? How should we use the answers from these questions and the feedback they give us to improve our science, debate, decisions, implementation, and monitoring? If we decide that we need to change some previous decisions and practices that aren't helping us to achieve our goals, what is the best way to phase them out and initiate more effective, justified decisions and practices?

It is self evident from the few questions posed here that a comprehensive, fair, and ultimately successful management process must include a broad range of participants—in fact, anyone who will in any way be affected by the decisions that are made. Management must be a careful, deliberative, lengthy exploration of logistics, ramifications, justifications, and other considerations. By asking these and numerous other questions, we can ensure that the institutional system of wildlife management takes advantage of the valuable learning opportunities that are currently available. We need to examine what we have done in the past, build on successes, and avoid repeating less successful practices (or continuing to do the same old thing even though circumstances have changed).

To repeat, the carnivore management process is really about people—what we believe and value, how we interact, and especially how we set up and carry out practices to limit harmful impacts on each other, on wildlife, and on the environment. Because this management process determines what happens to a public resource, it should incorporate the highest standards in decision making. It should be open, fair, comprehensive, reliable, creative, rational, integrative, effective, constructive, timely, dependable, independent of special interests, fully contextual, respectful, balanced, prompt, ameliorative, reputable, and honest.[8] These standards are ideals, and although they may never be perfectly achievable, we should still make every effort to meet them. They are designed to ensure that decisions serve the common interest; to strive to meet them is fundamental to democracy. The challenge for those who are concerned about large carnivore management is to take a hard look at whether these ideals are being met in the existing institutions of wildlife management, and whether they are even being pursued.

At present all kinds of special interests are competing to influence the carnivore management process. Rather than working toward their common interests, each of these interests is promoting its own definition of the problem and its own preferred solution. The concepts of "common interest" and "special interest" are familiar to most people.[9] Common interests are those widely shared within a community (e.g., having safe drinking water and a healthy ecosystem). Special interests are those that benefit only one group at the expense of others (e.g., poaching wildlife in a national park). There are many kinds of special interests, but all tend to mask their claims in the symbols and language of the common interest. It is not easy to sort out valid common interest demands from those of special interests, nor is it easy to determine which types of interests are presently served by the institutions of wildlife management. Nevertheless, it is only by sorting these things out that we will be able to find win-win solutions and learn how to coexist with one another and with large carnivores.

Practice-Based Learning

Grizzly bears, wolves, and mountain lions are the largest carnivores in the West. They are at the top of the food chain, are wide ranging in their use of habitats, and are considered to be important indicators of nature's responses to change. Predation by this suite of carnivores constitutes a powerful natural process in the ecosystems they occupy. As a result of years of scientific study, the ecology of these animals is becoming better known. For example, we now know more about their true predatory abilities. Recent research in the Greater Yellowstone region on these species, as well as coyotes, lynx, wolverines, and black bears, has contributed substantially to our understanding of natural history, ecological processes, and human impacts on ecosystems.[10]

Even so, we still do not know everything that we would like to know about large carnivores. Fortunately, we do not have to wait for a complete picture of the ecology of these animals in order to improve management, perhaps dramatically. Moreover, as we emphasize throughout this book, management is as much a response to people's beliefs about carnivores, especially about predatory behavior and potential danger to humans, as it is a response to the animals' actual ecology.

Focusing on People's Beliefs

Looking at people's beliefs is a logical place to start to understand the ongoing management process and how to make it more realistic and consistent with the actual characteristics and behavior of the animals. The different beliefs that people hold about large carnivores are tied closely to their basic beliefs about themselves, about appropriate relationships with nature, about the value and rights of individuals, and about how decisions should be made within their communities and the nation. Unfortunately, most people's beliefs about carnivores in the West today are based on anecdotal stories and second- or thirdhand experiences. Very few individuals have had firsthand experiences with carnivores, and although some of these experiences have been positive, some of the more sensational experiences have involved threats to personal or economic security. The diversity of views can be seen in regional newspapers and in books such as *Tales of the Grizzly* and *Tales of the Wolf.*[11]

Steven Kellert and Carter Smith, sociologists at Yale University's School of Forestry and Environmental Studies, have investigated human attitudes and values toward large mammals.[12] According to their research, national studies suggest that there is a nearly equal split between Americans who view the wolf with affection and admiration and those who view it with fear and antagonism. Also, people's beliefs vary by gender, geographical location, and socioeconomic group. Kellert and Smith found that women generally have stronger "humanistic" and "moralistic" values than men. Humanistic values deal with people's emotional affinity for nature. Moralistic values emphasize a sense of ethical or moral responsibility for conserving, protecting, and treating nature and animals well. Women tend to show greater affection and emotional attachment to large mammals, whereas men typically show stronger "utilitarian" values, emphasizing the practical and material importance of nature, and "dominionistic" values, showing an inclination to subdue and master nature. Farmers, ranchers, loggers, miners, and residents of open country tend to hold stronger utilitarian and dominionistic values than do other people.

More affluent people tend to show stronger "naturalistic" values, which focus on personal pleasure and satisfaction from direct experience and contact with animals in their natural habitats. College-educated people typically value large mammals for "naturalistic, humanistic, moralistic, and scientific" purposes more strongly than do people with just a high-school education, who tend to be more "utilitarian, dominionistic, and negativistic." Scientific values emphasize the empirical study and understanding of these animals and the ecosystems in which they live. Negativistic values express fears, anxieties, and awe about nature. Kellert and Smith observed that large mammals "represent far more than biological realities for people." They have "symbolic" values for everyone, even though these differ widely. The importance of these species "resides more in the peculiar human capacity to construct a world of meaning stretching far beyond the physical constraints of empirical reality."

Our knowledge of how people in Wyoming in the past valued large carnivores is very limited, although we do know that these animals were largely eliminated from the landscape. Today many people's beliefs in this region seem to be wrapped up in the frontier/cowboy myth, tied to convictions about material progress, individualism and states' rights, and a strong ideology about power relationships. This dominant belief system is largely dominionistic, utilitarian, and negativistic, to use Kellert's value classification. However, other people who are involved with carnivore management, operating under different belief systems, emphasize moralistic, humanistic, and naturalistic values. These values and symbolic meanings are little studied or appreciated in most places.

People's beliefs and symbols help them to sort through their experience and what they view as "facts" to find meaning. When people don't know the facts they fill in the gaps with images and symbols. Factual information about large carnivores in the region is difficult to get, often misused, and actively contested. Yet the limited factual information that is available figures directly into how people understand and participate in the management process. Perceived facts, beliefs, and symbols are mixed together in thought, advocacy, and interest-based politics, contributing to the complexity and contentiousness of carnivore management.

Focusing on Management

A variety of federal and state agencies hold management authority for grizzly bears, wolves, and mountain lions. No one agency has complete control; instead management exists in a complex web of interactions.[13] For grizzlies, the U.S. Fish and Wildlife Service has lead responsibility because the species is listed as threatened under the federal Endangered Species Act. However, much of the management of grizzlies in the state is actually carried out by the Wyoming Game and

Fish Department. The U.S. Forest Service, especially in its dealings with livestock permitees, loggers, and recreational users, also manages grizzly bears. The U.S. Fish and Wildlife Service, the National Park Service, and the states all conduct research on bears. In short, the management arena for grizzlies is highly segmented.

For wolves, the story is different. The species has only recently been downlisted from endangered to threatened under the Endangered Species Act, and the U.S. Fish and Wildlife Service conducts all management. The state of Wyoming—by its own choice—has no direct involvement, although this may change in the future.

For mountain lions, the situation is different again. Because of the larger numbers of mountain lions, they are not listed under the Endangered Species Act. The state has exclusive management authority.

Many other participants are involved in, affected by, or interested in management of these large carnivores, including environmental organizations, the oil and gas industry, loggers, miners, recreationists, scientists, ranchers, governments, and other state and federal agencies (such as the Wyoming Department of Transportation and the federal Bureau of Land Management). As this brief overview shows, authority and control are highly fragmented in the large carnivore management arena, which makes it especially challenging to establish and run an effective overall institutional system for management.

We know quite a bit about these three species in Yellowstone National Park and the immediate surrounding area. It is unusual to have this amount of information for an entire "guild" or "suite" of carnivores, so this area is quite unique in this regard. But in the southern reaches of western Wyoming we are not so fortunate. There, available information about the status of these animals and the direction and progress of management programs is widely scattered in federal and state management plans, annual reports, reports of public meetings, scientific and management papers, public education documents, and Web sites. Although there is a variety of possible sources, information is not always easy to come by, nor is it always reliable. Moreover, special interest groups sometimes use, emphasize, and distribute information in selective ways that serve their own partisan demands. This makes it very difficult to find out what actually is happening, and thus it is nearly impossible for anyone to make an independent assessment of the adequacy of management programs. This is especially the case for grizzly bears.

Because of all the factors that make carnivore management so complex, we must learn to use the information we already have, together with new information as it becomes available, and the vast experience of ranchers, wildlife biologists, and others, to craft better ways of doing things. We have a monumental task ahead of us: integrating authority and control as well as information and experience into an effective, efficient, and equitable institutional system for managing large carnivores and other wildlife. To determine where best to begin this task, we need to develop a better understanding of the problem of large carnivore management.

Being Problem Oriented

How we define a problem determines whether we can solve it or not. There are good problem definitions, which help people in their search for effective, efficient, and equitable solutions, and bad problem definitions, which encourage "partial solutions" or "the right solution for the wrong problem." Bad problem definitions mislead people, confuse outcomes, discourage participants, and foster chronic conflict. The trick is to cultivate the ability to distinguish a good problem definition from a bad one. Creating sound solutions to large carnivore management must begin with sifting through the competing problem definitions now in play so that we can define the issues realistically and comprehensively.[14]

The tried-and-true way to tease out the best problem definition is to be problem oriented in our inquiry. This is key to a practice-based learning approach. Systematically examining what we have actually done in the past and what the results have been will give us the most information (and the most practically useful information) about how we can do better in the future. To conduct such a systematic examination we must ask five basic questions: (1) What are the goals of large carnivore management? (2) What are the achievements and shortfalls to date with respect to these goals? (3) What forces, factors, and conditions influence how management works and lead to these achievements and shortfalls? (4) Is management likely to continue working this way in the future or will it change; if it will change, in what ways? The answers to these first four questions will give us real insight into how well the present system of wildlife management is working. If we find out that it is not working well and our goals are not being achieved, then a problem exists. This leads us to ask the final question: (5) Based on the information we have learned from our inquiry, what options or alternatives do we have to solve the problems that we have identified, by changing the factors that are causing them? Or, operationally, how can we improve or upgrade the present management process to overcome problems?

What Are the Goals of Large Carnivore Management?

We can judge the success of the present management process only if we know what we are trying to achieve. In other words, we must state our goals so that we can measure actual performance against them. We may decide that the goal is to live in coexistence with large carnivores, or to keep carnivores at the lowest possible densities and distribution, or to keep bears and wolves confined to Yellowstone National Park, or some other state of affairs. In general terms the overall goal is always to clarify, meet, and sustain our common interests, but the discussions, debate, information gathering, advocacy, judgments, and wrangling that must take place among all interested people before we come to a workable agreement on more specific goals can be difficult and lengthy.

In the meantime we can take the goals spelled out in existing federal and state management plans for grizzlies, wolves, and mountain lions as "working specifications" of the community's common interests. These plans give specific objectives for managing the three species along with a more encompassing (though often not explicitly stated) goal of *restoring large carnivores and carrying out programs for coexisting with them in ways that will engage the public and benefit from public support.* This overall goal, which certainly appears to be a common interest goal that serves deliberative democracy and a healthy environment, is likely to be supported by all or most people, at least in principle. In fact, this goal is now supported broadly in national, state, and local policy.

Comparing this goal to actual achievements, we can see that the performance of present management is decidedly mixed. There are some successes and some failures, as the cases in this volume show. Only time will tell whether this goal really serves common interests and whether the means we have chosen to implement it are effective or not. In the meantime, though, we can still ask three questions about the process itself to determine if it at least is working in the common interest: Is it inclusive and open to broad participation? Does it meet the valid expectations of the participants (based on facts)? As management is implemented or practically tested, is it responsive and adaptable when errors occur or when the context changes?[15]

What Are the Achievements and Shortfalls to Date?

Although the other chapters in this volume examine trends in more detail, one conspicuous failing of the current management process is that it often seems to serve special interests at the expense of common interests. In particular, the overriding goal of carnivore management is commonly subordinated to the age-old power politics of federal versus states' rights, with some interests seeking to achieve their personal ends by promoting state power, while others turn to the federal level to impose alternative agendas. The repeated failure to identify and pursue the common interest is substantiated by many letters people have written to agencies and the press and by numerous reviews of current management processes (some of which have been cited earlier).

The failure is also demonstrated by people's responses to management. For example, some people have become fierce advocates for special interests, some have abandoned the public process and formed their own decision-making groups, and some, despairing that anything can be done to improve matters, have dropped out altogether. Others have gone to the media to pressure government to be more fair, open, and common-interest-oriented, and still others have turned to the courts for redress. This overall pattern of participation (or lack thereof) leads to heightened conflict, and recently, the conflict appears to be intensifying. The federal and state governments largely structure the management process and control how it unfolds. They largely determine who participates, how, and with what outcomes. Consequently, it is the federal and state governments that are responsible for much of the current state of affairs.

Why Does Management Work the Way It Does?

The many forces, factors, and conditions that influence human-carnivore management problems are not aligned in a simple cause-and-effect chain. They are instead complexly interconnected in synergistic ways, perhaps far beyond human understanding. Among the more far-removed factors are global forces causing large-scale changes, such as the development of global markets and national and international political realignments. These forces influence local beef prices, for example. The more drastic the change, the more people struggle to adapt, and the more firmly they cling to their basic worldviews. Personal meaning, dignity, and feelings of empowerment are rooted in these worldviews. Public and personal reaction takes the form of resistance to agencies, government, and anyone who is perceived as the "opposition," and even resistance to change in general. Agencies come to be seen as the enemy, associated with perceived losses of freedom and self determination and with too much top-down regulation. Outsiders, even those in adjacent communities, become the enemy too, especially if they hold different beliefs and values.

There are many reasons why the management process is problematic. Chief among them are (1) the way the process is presently understood and the ways in which people participate in it, (2) the diversity of beliefs, values, and knowledge that people hold, which are often in dramatic opposition to one another, and (3) the structure, or very "backbone," of the institutional system of wildlife management, which significantly limits the discourse through which people interact with one another and resolve their differences. Often this discourse is not as reasonable, practical, justified, or respectful as it could or should be, and it does not lead to the buildup of social capital and trust that is desperately needed.[16] All these factors and others come together to make up the present problem of large carnivore management.

In addition, the management process is problematic because of other contextual issues. Wyoming is undergoing many rapid changes in demographics, economics, and land uses, including important changes in ranching and recreational activities. Carnivores themselves are increasing in abundance and distribution. Also, there are few institutions in the region to help people (on their own terms) to develop effective community problem-solving mechanisms for win-win or integrated solutions. Useful examples of people successfully engaging in deliberative democracy and genuine problem solving are rare, so there are few models for people to follow as they attempt to resolve difficult management issues.

Management agencies are trying to keep up with changes in the region and the public's diverse demands, but overall, government has not been fully responsive or adaptive to these trends in an "active learning" sense. Government often over-relies on its own authority, the standard operating procedures of bureaucracy, and the power of traditional agency professionalism and expertise to understand and solve problems. This narrow, traditional approach is seen as heavy handed and coercive by those on the receiving end; it excludes and alienates people, regardless of where they stand on the carnivore issue. Generally, the organizational arrangements, cultures, and leadership of the agencies are not keeping up with the pace and size of the growing problems in carnivore management, conservation, and natural resource management policy.

How Will the Carnivore Management Process Likely Work in the Future?

The most likely scenario is that the present management process will continue as is unless someone proactively intervenes to make it work better. Change will require leaders who are willing to take risks. If the current situation continues, things may get much worse as wolves and grizzlies expand their densities and range and come into contact with extensive sheep allotments farther south on national Forest Service and Bureau of Land Management lands. Predation on livestock could increase substantially. Furthermore, when the state of Wyoming assumes responsibility for wolves and perhaps grizzly bears, which will occur when these species are removed from Endangered Species Act protection, it will be moving from the frying pan directly into the fire as it is called on by interest groups to kill more "problem" animals. This will no doubt bring the Wyoming Game and Fish Department into acute conflict with the national public, which strongly supports conservation and protection of these species.

Alternatives Open to Us

The answers to the four questions in the "Being Problem Oriented" section provide the information necessary to answer our fifth and final question, What options or alternatives do we have to solve the problems we have identified? Despite some obvious successes, carnivore management remains inadequate in many ways. Getting people and predators to coexist sustainably has proven to be extremely difficult.

People have not paid enough attention to the kinds of knowledge that are needed, they have not measured the quality of the present management process itself, and finally, the process has not been as inclusive and collaborative as it should be. In short, the institutional system of wildlife management has performed suboptimally in many ways. The challenge or problem, as defined here, seems self evident to a growing number of people. From a substantive point of view, minimizing the damage that carnivores cause and changing what carnivores mean to people are the two immediate targets. The first can be accomplished relatively easily, in comparison with the complexities of changing the meaning and symbolic politics involved, but in both cases we must be adaptive, develop leadership, and learn through practice-based work on the ground.

Given the information gathered in answer to the above questions, then, what are the obstacles that stand in the way of achieving the goals we agree on and what can be done to overcome these obstacles? As developed in later chapters, we need to follow through on the following strategies.

Emphasize Bottom-up Rather Than Top-down Solutions

Small-scale prototypes, or trial interventions, can be developed with the cooperation and participation of ranchers, environmentalists, business, and government, to test new approaches and methods designed to reduce carnivore damage and change carnivore meanings. Prototypes that work can be replicated or adapted in other settings. Those that fail can be shut down without great cost. Such a prototyping strategy can build a record of successes without much risk and significantly improve the management process in the short term. Prototyping is a proven strategy to address species and ecosystem management problems.[17] Several chapters in this volume elaborate on this strategy and describe approaches and methods that could be used.

Build on Past Successes

There is no need to reinvent the wheel to improve carnivore management. Considerable experience has been gained in other states and in Canada from living with, studying, and managing these three species. Both ranchers and carnivore biologists know a great deal about minimizing livestock losses to carnivore predation. Getting these groups to work together to reduce losses and other conflicts is one obvious way to improve management. Examples of this kind of practical work in the field include the efforts of Mark Bruscino (Wyoming Game and Fish Department), Mike Jimenez (U.S. Fish and Wildlife Service), Barb Franklin (U.S. Forest Service), Timm Kaminski (Mountain Livestock Cooperative) and Steve Primm and Seth Wilson (Northern Rockies Conservation Cooperative) and their associates. Some of these efforts are described in the following chapters. Many ranchers are currently working cooperatively with government and conservation groups to improve matters on the ground. Harvesting this vast experience, building on it, and putting it to work everywhere is one major way of moving forward toward more successful coexistence.

Improve the Institutional System for Managing Wildlife

The region already has extraordinary resources to address the institutional challenges of carnivore management—wealth, community commitment, extensive protected areas on federal lands, wildlife that draws visitors from around the world, a powerful set of government agencies and conservation organizations, and a generally conservation-minded public. The challenge here is to identify what additional institutional innovations could improve decision-making processes and secure regionwide coexistence of large carnivores and human communities. This high-profile region could be an institutional exemplar, setting the standards for effective problem solving and democratic governance for wildlife conservation. This can be accomplished by implementing the first two strategies outlined above and by bringing in a new kind of "transformative" leader who is willing to take risks, work hard, and—like a "maestro"—help people to achieve their common goals. Existing managers must also become better problem solvers and will need training and other assistance to develop the necessary skills. Chapter 7 is dedicated to the institutional system of wildlife management and examines these and other ways to make it more effective.

Points of leverage exist where well-designed interventions could improve the process for managing wolves, grizzly bears, and mountain lions. We must determine the most effective levers and apply them in the three strategies described above. Each chapter in this volume offers recommendations to accomplish this.

There are many reasons why the management process should be upgraded to bring about these changes. First, people clearly have a shared interest in resolving this matter. Second, we already know how to make things better for both people and carnivores, but have not been making adequate use of this knowledge. Third, there are successful efforts underway in the field today that could be adapted, expanded, and copied.

Conclusion

For decades wolves were absent from western Wyoming, and grizzly bear and mountain lion populations were very low. All have returned in recent years. The overall goal is to restore and coexist with these animals as part of viable regional populations that benefit from broad public support. State and federal legislation is in place or is being developed that supports this goal, but it needs greater specificity, refinement, and practicality. Much but not all management policy in recent years also supports this goal, as do many people regionally and nationally. However, there are also people, organized interests, and agencies that do not agree with this goal and have actively worked against it, and many human practices do not support it (e.g., leaving food and garbage available to bears, letting children play unsupervised in mountain lion habitat, leaving livestock unattended). The present conventional problem-solving approaches, inefficient government organizations, and local governance mechanisms have failed to find a workable balance for human-carnivore coexistence. Yet both carnivore (especially wolf) and human densities are expected to increase in the future, bringing them into closer and more frequent contact. Most of our contacts with large carnivores are benign, but a few result in conflicts with human activities or risks to human safety.

Lack of a fully effective management process leads to many other problems and considerations. There is considerable uncertainty and even strong disagreement about the viability of these species' populations, the conservation measures needed, and the kind and degree of public support present or needed. Some people argue that we need to know a lot more about the ecology of these species, about public attitudes, and about the effectiveness of management actions before we can establish successful management programs. Other people argue that policy goals are too vague and need more realistic specificity to be practical. Some people feel that implementation of existing management is biased against these animals and toward historic land uses and traditional, local, special interests. Some feel that effective means to resolve disputes among different special interests are lacking and that relying on governmental power and the courts is a coercive rather than constructive way to settle differences. Finally, transferring management from one government agency at the federal level to another at the state level, as called for by the plans to delist threatened and endangered carnivores, is new ground that is full of pitfalls, including the possibility of endless conflict and failure. This transition must be considered carefully to make it work well.

The present management process and the structure of the institutional system of wildlife management also bring out other larger questions about the adequacy of leadership, capabilities of agencies, and sufficiency of resources. These and other problems must be defined accurately and resolved successfully if the goal of conserving and coexisting with large carnivores is to be achieved.

We hope that our findings, detailed in subsequent chapters, will stimulate discussion and action, and we hope that our recommendations are sufficiently pragmatic to open new possibilities for collaboration and cooperative, creative problem solving.

References

1. These diverse meanings are discussed in the following publications, as is the difficulty of restoring these species and their ecosystems: R. P. Reading and B. J. Miller, eds., 2000, *Endangered animals: A reference guide to conflicting issues,* Greenwood Press, Westport, CT; T. W. Clark, R. P. Reading, and A. L. Clarke, eds., 1994, *Endangered species recovery: Finding the lessons, improving the process,* Island Press, Washington, D.C. Differences in meaning are also clear in articles from Jackson, Wyoming, newspapers, such as R. Huntington, 2002, "Game and Fish working to trap grizzly family," *Jackson Hole Guide,* Aug. 28, A11; R. Huntington, 2002, "Park confirms first cattle kill by wolves," *Jackson Hole Guide,* Aug. 28, A2; J. Goodall, 2002, "Mountain lions need protection by law," *Jackson Hole News,* June 26, 5A.

2. C. Urbigkit, 2002, "The coup counties: Western Wyoming commissioners outlaw grizzlies and wolves," *Range,* Fall, 30–31; T. Wilkinson, 2002, "Hatred toward wolves a sign of bigger issues," *Jackson Hole News,* Aug. 21, 5A.

3. T. McNamee, 2003, "Tinkering with nature," *High Country News* 35(6), 9; R. M. Ferris, M. Shaffer, N. Fascione, H. Pellet, and M. Senatore, 1999, "Places for wolves: A blueprint for restoration and long-term recovery in the lower 48 states," Defenders of Wildlife, http://www.defenders.org/pubs/pfw01.html; G. Hanscom, 2003, "The best restoration tools are fangs and claws," *High Country News* 35(6), 2.

4. These few articles are representative of hundreds over the last decade that clearly demonstrate the symbolic importance of large carnivores, which serve as surrogates for many other issues: T. Finley, 2002, "The real American wolfman," *Range,* Fall, 34–36; G. Ochenski, 2002, "A love story: Wyoming's Twin Creek Ranch embraces diversity," *Range,* Fall, 14–15, 19; B. Flanning, 2002, "Running wild in Yellowstone: The magnificent elk herds of Montana are diminishing due to flourishing flesh-tearing predators; Something's got to change," *Range,* Fall, 18–19; T. C. Tabor, 2002, "The big track of the cat," *Range,* Fall, 21.

5. The symbolic politics, as well as how to improve matters practically, have been written about extensively. For example, see S. A. Primm, 2000, "Real bears, symbol bears, and problem solving," *NRCC News* (Northern Rockies Conservation Cooperative, Jackson, WY) 13, 6–8; S. A. Primm, 2001, "Participatory processes for better problem solving," *NRCC News* 14, 10–11; S. A. Primm, 2002, "Grizzly conservation and adaptive management, *NRCC News* 15, 7–8; T. Kaminski, C. Mamo, and C. Callaghan, 2002, "Managing wolves and bears where private ranches meet public reserves," *NRCC News* 15, 11–12. See also S. A. Primm, 1996, "A pragmatic approach to grizzly bear conservation," *Conservation Biology* 10, 1,025–1,033; S. A. Primm and T. W. Clark, 1996, "Making sense of the policy process for carnivore conservation," *Conservation Biology* 10, 1,035–1,045.

6. The human tendency to divide people into "us vs. them" has been extensively studied. See, for example, E. J. Sahurie, 1992, *The international law of Antarctica,* New Haven Press, New Haven; A. Flores and T. W. Clark, 2001, "Finding common ground in biological conservation: Beyond the anthropocentric vs. biocentric controversy," 241–252 in T. W. Clark, M. Stevenson, K. Ziegelmayer, and M.R. Rutherford, eds., *Species and ecosystem conservation: An interdisciplinary approach,* Yale University School of Forestry and Environmental Studies Bulletin No. 105.

7. This definition of an institution is from *Webster's New Universal Unabridged Dictionary,* 1994, Barnes & Noble, New York.

8. These standards were first recommended by Harold Lasswell, one of the most creative and productive social scientists of the last century, in his 1971 book, *A pre-view of policy sciences,* American Elsevier, New York. They are also discussed in T. W. Clark, 2002, *The policy process: A practical guide for natural resource professionals,* Yale University Press, New Haven.

9. Two books with good discussions of common and special interests are R. A.

Dahl, 2000, *On democracy,* Yale University Press, New Haven; and R. A. Dahl, 1970, *After the revolution?* Yale University Press, New Haven. Both of these books serve as background to Lasswell (1971) (see endnote 8 above). Lasswell (1971) and Clark (2002) both discuss five additional standards for the most important decisions, or "constitutive" management processes, including ensuring that common interests will prevail over special interests.

10. The taxonomy, history, status, ecology, and behavior of these carnivores in Wyoming and in the Yellowstone region are described in T. W. Clark and M. R. Stromberg, 1987, *Mammals in Wyoming,* University of Kansas Press, Lawrence; and T. W. Clark, A. P. Curlee, S. C. Minta, and P. M. Kareiva, eds., 2000, *Carnivores in ecosystems: The Yellowstone experience,* Yale University Press, New Haven; T. W. Clark, P. C. Paquet, and A. P. Curlee, eds., 1996, "Large carnivore conservation in the Rocky Mountains of the United States and Canada," *Conservation Biology* 10, 936–1058.

The following papers summarize what is known about the role of predation in ecological systems: J. A. Estes, K. Crooks, and R. Holt, 2001, "Predation and diversity," 857–878 in S. Levin, ed., *Encyclopedia of biodiversity,* Academic Press, San Diego; B. Miller, B. Dugelby, D. Foreman, C. Marinez del Río, R. Noss, M. Phillips, R. Reading, M. E. Soulé, J. Terborgh, and L. Willcox, 2001, "The importance of large carnivores to healthy ecosystems," *Endangered Species Update* 18, 202–210; L. Oksanen and T. Oksanen, 2000, "The logic and realism of the hypothesis of exploitation ecosystems," *American Naturalist* 155, 703–723; T. Oksanen, L. Oksanen, M. Schneider, and M. Aunapuu, 2001, "Regulation, cycles and stability in northern carnivore-herbivore systems: Back to first principles," *Oikos* 94, 101–117; and O. J. Schmitz, P. A. Hamback, and A. P. Beckerman, 2000, "Trophic cascades in terrestrial systems: A review of the effects of carnivore removals on plants," *American Naturalist* 155, 141–153. For an excellent review of Yellowstone's carnivore research, trends in research, how carnivores best inform ecological concepts, and the future of carnivore research (with an eye toward improving carnivore management policy), see S. C. Minta, P. M. Kareiva, and A. P. Curlee, 1999, "Carnivore research and conservation: Learning from history and theory," 323–404 in T. W. Clark, A. P. Curlee, S. C. Minta, and P. M. Kareiva, eds., *Carnivores in ecosystems: The Yellowstone experience,* Yale University Press, New Haven.

There is, of course, growing interest for restoring and protecting large carnivores. See, for instance, D. S. Maehr, R. F. Noss, and J. L. Larkin, eds., 2001, *Large mammal restoration: Ecological and sociological challenges in the 21st century,* Island Press, Washington, D.C.

11. D. Casey and T. W. Clark, 1996, *Tales of the wolf: Fifty-one stories of wolf encounters in the wild,* Homestead Press, Moose, WY; T. W. Clark and D. Casey, 1992, *Tales of the grizzly: Thirty-nine stories of grizzly bear encounters in the wilderness,* Homestead Press, Moose, WY. There are many other such books in print.

12. See S. R. Kellert and C. P. Smith, 2000, "Human values toward large mammals," 38–63 in S. Demarais and P. R. Krajusman, eds., *Ecology and management of large mammals in North America,* Prentice Hall, Upper Saddle River, NJ. See also S. R.

Kellert, 1996, *The value of life: Biological diversity and human society,* Island Press, Washington, D.C.; and S. R. Kellert, 1997, *Kinship to mastery: Biophilia in human evolution and development,* Island Press, Washington, D.C.

13. A chief feature in this complexity, beyond the cultures involved, is the battle for decision-making power between the federal and state governments. The national-level focus of the federal agencies often clashes with the states' rights ideology of the state of Wyoming. This federal-state clash has existed since the founding of the republic; it led to the Civil War (which decided the issues in favor of federalism), and it is alive and well today in places like Wyoming, where most decisions about managing natural resources turn into bitter struggles between federal and states' rights. One example of how this power contest hindered both species conservation and the establishment of a cooperative program was the black-footed ferret case in Wyoming; see T. W. Clark, 1997, *Averting extinction: Reconstructing endangered species recovery,* Yale University Press, New Haven. Other reading on this subject includes: D. Dary, 2002, *Cowboy culture: A saga of five centuries,* University of Kansas Press, Lawrence; F. McDonald, 2002, *States' rights and the union,* University of Kansas Press, Lawrence; and R. V. Percival, A. S. Miller, C. H. Schroeder, and J. P. Leape, 2000, *Environmental regulation: Law, science, and policy,* Aspen Law & Business, New York.

14. The importance of problem definition is often overlooked in natural resource management. Perhaps the most important paper on this topic is J. A. Weiss, 1989, "The powers of problem definition: The case of government paper work," *Policy Sciences* 22, 97–121. Other writings emphasize the import of realistic problem definition and how definitions are constructed. For example, see S. A. Primm and T. W. Clark, 1996, "The Greater Yellowstone policy debate: What is the policy problem?" *Policy Sciences* 29, 137–166; T. W. Clark, A. P. Curlee, and R. P. Reading, 1996, "Crafting effective solutions to the large carnivore conservation problem," *Conservation Biology* 10, 940–948; J. L. Scheuer and T. W. Clark, 2001, "Conserving biodiversity in Hawai'i: What is the policy problem?" 159–184 in T. W. Clark, M. Stevenson, K. Ziegelmayer, and M. Rutherford, eds., 2001, *Species and ecosystem conservation: An interdisciplinary approach,* Yale School of Forestry and Environmental Studies Bulletin No. 105; T. W. Clark, D. J. Mattson, R. P. Reading, and B. J. Miller, 2001, "Interdisciplinary problem solving in carnivore conservation: An introduction," 223–240 in J. Gittleman et al., eds., *Carnivore conservation,* Cambridge University Press, Cambridge.

15. See Cromley, 2002, *Beyond boundaries: Learning from bison management in Greater Yellowstone,* Ph.D. dissertation, School of Forestry and Environmental Studies, Yale University, New Haven.

16. The idea of "social capital" as a foundation for solving common interest problems is described in R. D. Putnam, 2000, *Bowling alone: The collapse and revival of American community,* Simon and Schuster, New York. An excellent overview and set of cases that illustrate what is involved in finding common ground is R. D. Brunner, C. H. Colburn, C. Cromley, R. A. Klein, and E. A. Olson, eds., 2002, *Finding common ground: Governance and natural resources in the American West,* Yale University Press, New Haven.

This book provides a good account of what is involved in good management in the common interest; it also uses cases on wolf and bison management from the Yellowstone region as examples. Finally, "bridges" that must be built to overcome "barriers" to common interest outcomes are described in T. W. Clark and S.C. Minta, 1994, *Greater Yellowstone's future: Prospects for ecosystem science, management, and policy,* Homestead Publishing, Moose, WY. See also L. H. Gunderson, C. S. Holling, and S. S. Light, eds., 1995, *Barriers and bridges to the renewal of ecosystems and institutions,* Columbia University Press, New York.

17. Prototyping, or the innovation strategy, is widely used in actual management. See R. D. Brunner and T. W. Clark, 1997, "A practice-based approach to ecosystem management," *Conservation Biology* 11, 48–58; T. W. Clark, G. N. Backhouse, and R. P. Reading, 1995, "Prototyping in endangered species recovery programs: The eastern barred bandicoot experience," 50–62 in A. Bennett, G. Backhouse, and T. Clark, eds., *People and nature conservation: Perspectives on private land use and endangered species recovery,* Transactions of the Royal Zoological Society of New South Wales, Mosman; T. W. Clark, R. P. Reading, and G. N. Backhouse, 2002, "Prototyping for successful conservation: The eastern barred bandicoot program," *Endangered Species Update* 19(4), 125–129.

Chapter **2**

Management Context: People, Animals, and Institutions

Dylan Taylor and Tim W. Clark

Successfully managing large carnivores—mountain lions, wolves, and grizzly bears—means sustaining viable animal populations and vibrant human communities at the same time. This requires a thorough understanding of the context in which the management process is carried out, including human values and practices, and the requirements of the animals.[1] Context matters enormously. Regrettably, managers often neglect to obtain adequate contextual information and use it in decision making. This leads to failures when overlooked or misconstrued aspects of the context end up playing significant roles in determining management outcomes.

"Context" is the set of conditions that shapes both problems and solutions.[2] It includes the people, groups, and organizations who are involved or affected, their perspectives (including their beliefs and demands), the ecological and other features of the situation, values at stake, strategies used, outcomes sought, and longer term effects. In any context, people typically make claims of one kind or another, which are often met with counterclaims by those who hold different perspectives. Reconciling such competing claims in the common interest is the principal task of a good management process.

Setting the stage for subsequent chapters, this chapter describes the context for large carnivore management in the region covered by this volume (western Wyoming south of Yellowstone National Park—figure 2.1). It examines the landscape, wildlife, people and their culture,

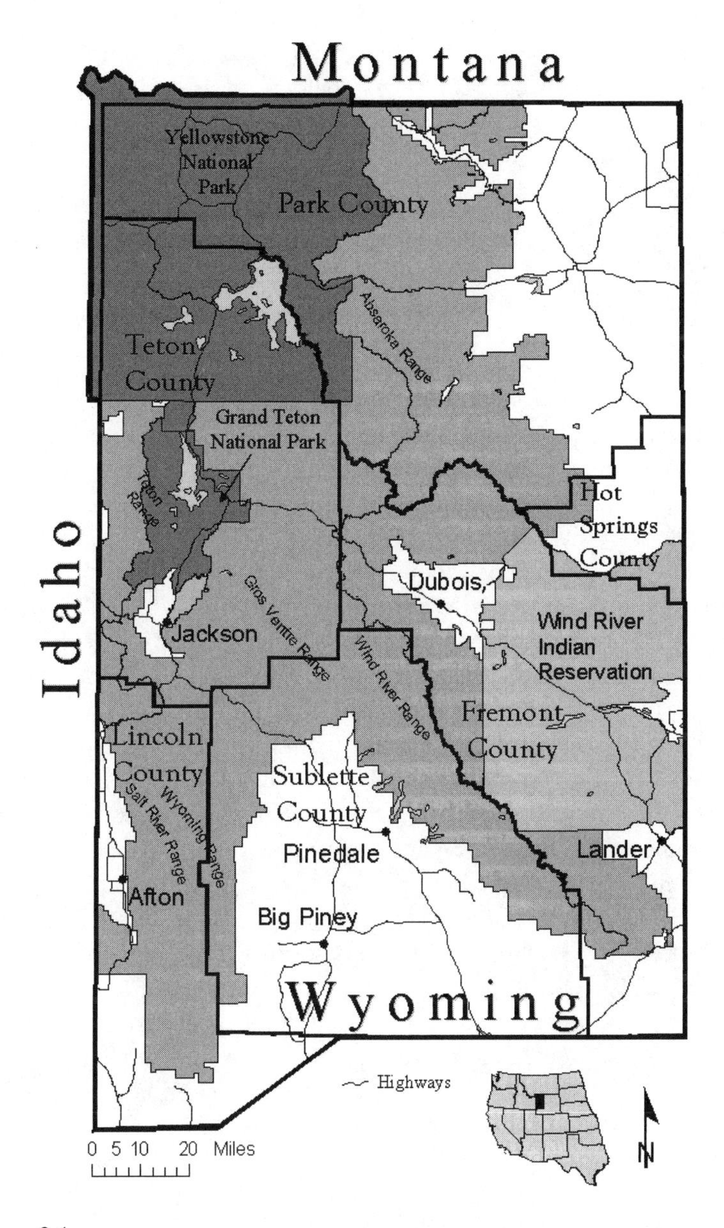

Figure 2.1
Map of the southern part of Greater Yellowstone showing location of key features such as counties, towns, highways, and mountains.

and institutional arrangements. It analyzes key trends, their causes, and the implications for future relations. Finally, it offers options to improve decision making for the coexistence and sustainability of carnivore and human communities. The options focus on improving contextual understanding, which will guide the entire management process toward the common interest.

In the context of western Wyoming, the choices that managers have made about wildlife and land management have often turned out to be highly contentious. Individuals, groups, and organizations with competing values have wrestled in court, public meetings, and other forums, each seeking the outcome that best suits their own interests. Powerful special interests have made it difficult to clarify, secure, and sustain the common interest. The resulting conflict has been corrosive to the management process and to social capital. People are finding it more and more difficult to work together to achieve acceptable outcomes, and trust in government and agency managers is plummeting. This is a highly problematic context in which to achieve successful outcomes, especially concerning large carnivores.

The features of this context must be "mapped" and understood as a basis for more successful management. Such a map is not currently being used in carnivore management, although established methods for comprehensively and realistically mapping contexts are well known. The contextual map sketched out in this chapter is a beginning, which could be refined and expanded to become a reliable reference that would greatly help managers and all other participants in carnivore management. The map should be continually updated with input from citizens, experts, and others. Competing interests should be involved in its making to the maximum extent practicable, and the map itself should reflect those various interests. In this way, it might reveal more agreement, or common ground, than first appearances suggest and could give people a footing on which to build a cooperative program for coexisting with carnivores.

The Setting—Landscape, People, and Wildlife

Western Wyoming is indeed a beautiful place. Its vistas are breathtaking. Its mountains and basins, rivers and streams, forests and deserts make up a landscape unparalleled in grandeur. It has abundant wildlife and natural resources (such as timber, minerals, and natural gas), and a rich human history. This landscape is the setting for the carnivore management drama examined in this volume. It contains a cast of characters, both human and animal. Culture and the history of land use are central to the story. This system behaves as it does in part because of the institutions of wildlife management that are in place, which set the conditions and structure how people interact. These institutional arrangements, which we call the "institutional system of wildlife management," are the backbone of the people-carnivore dynamic.

Landscape and Physical Setting

Western Wyoming is a major arena for large carnivore management.[3] It is a matrix of federal, state, and privately owned lands, approximately 22,000 square miles in size, including Grand Teton National Park, National Elk Refuge, and Bridger-Teton National Forest. Ecologically, this region contains the southern portion of the Greater Yellowstone Ecosystem and adjacent lands; some consider it the southernmost part of the Yellowstone-to-Yukon bioregion. The Bridger-Teton National Forest alone is one of the largest national forests in the United States at 3.4 million acres, including more than 1.2 million acres of designated wilderness areas—land that is high quality habitat for wildlife. In human terms, the region includes the four Wyoming counties south of Yellowstone National Park—Sublette, Fremont, Lincoln, and Teton (figure 2.1). The main communities are Dubois (pop. 1,000), Pinedale (pop. 1,500), Afton (pop. 1,700), and Jackson (pop. 8,600).

The climate is characterized by long, cold winters and short, cool summers with average temperatures ranging from 16°F in January to 61°F in July. Snowfall, which can be substantial in the mountains, is the source for much of the region's water supply. Snowmelt recharges the aquifers and feeds streams and rivers throughout the summer.

The Gros Ventre, Teton, Salt River, Wind River, and Wyoming mountain ranges dominate the topography, rising to heights of 13,000 feet (figure 2.1). Forests of conifers such as lodgepole pine, Douglas fir, and Engelmann spruce cover the mountains and higher elevations. Aspens grow on many slopes. Open valleys of sagebrush, grasslands, and agricultural fields lie below the mountain ranges.

This region is known around the world for its spectacular scenery and diversity of wildlife. As part of what has been described as one of the "last large, nearly intact ecosystems in the northern temperate zone of the earth," it is considered "one of the world's foremost natural laboratories in landscape ecology and geology."[4] It is within this arena that people and carnivores interact in a search for coexistence.

People, Perspectives, and Culture

Humans dominate the landscape. Their practices have modified the physical and biological setting considerably. Current trends promise more, perhaps drastic, changes to the land, wildlife, and people in the future.[5] Understanding the human participants is essential to understanding what has happened, why, and what is likely to happen.

Human Participants

Many individuals and groups have a stake in carnivore management. Participants include people from all social groups and economic levels. These people have local, regional, national, and even international identifications. Their interactions drive carnivore management policy.

There are fewer people in Wyoming than in any other state. The total population of Sublette, Fremont, Lincoln, and Teton counties was 74,548 in 2000, just 1.5 times that of the city of Cheyenne. These four counties have population densities well below the state average of five persons per square mile, which is far below the national average of 80 persons per square mile. Only Alaska has a lower average population density than Wyoming. However, the population of western Wyoming has been growing greatly in recent decades. From 1990 to 2000 Teton County's population grew by 63 percent, while Sublette, Lincoln, and Fremont counties' populations grew by 22, 15, and 6 percent, respectively. This growth increases the pressure on both human and natural systems.

Perspectives and Values

Although most people in Wyoming value living with wildlife, their values differ greatly in terms of what is an appropriate level of coexistence and how to reach that level. The participants involved in large carnivore management in this region generally fall into three main groups, whom we call *localists* or the Old West, *environmentalists* or the New West, and *agency personnel*—each with its own viewpoint. We adopted these labels following Kimberly Byrd of the University of Minnesota, who used them to describe groups of participants in her study of wolf management in Minnesota.[6] Recent research in the Northern Rocky Mountains shows very similar groupings, so it seems helpful to apply Byrd's generalized profiles here.[7] Although they closely capture the main viewpoints of participants in western Wyoming, they are rough categories and thus are not perfect for describing everyone involved. Nevertheless, they provide a useful and revealing shorthand for referring to and understanding the dominant perspectives among the many and diverse participants. Both the Old West and the New West consider themselves stewards of this land. However, there is often a dramatic dichotomy in what these two groups value, in what they believe, in how they want the land and wildlife to be managed, and in fact, in what they do.

Localists, or the Old West, are those who identify with the traditional values and culture of old Wyoming. Some of these folks are descendents of families that settled the West in the late 1800s and early 1900s, who adopted rugged individualism, independence, and a strong sense of how to endure in a harsh environment. Others are not from original Wyoming families, but they identify with and share their "frontier/cowboy" values.[8] Generally, these people take pride in making a living from the land or being associated with this lifestyle, especially given Wyoming's dry summers and long, cold winters. Everything from license plates to the signs people plant in their front yards demonstrates that many people are proud of this heritage and image. It is reflected in the dress, customs, and identities people form and hold for themselves, as well as in their political expectations and demands.

Localists dominate the culture of the region and are present in all four counties. Generally comprising ranchers, politicians, conservationists, hunters, and outfitters, all living in the region, they control or influence most of the politics in the region through county commissions and in other ways. Their perspectives and value outlooks translate into actions that have affected the land and its wildlife dramatically since before statehood. Their use of resources is predominantly utilitarian (emphasizing the material benefits to be derived from nature) and dominionistic (concerned with dominating or controlling nature and animals), and their outlook results in practices that are often at odds with sustainable conservation. Today, these people are strongly tied to a philosophy of individual rights, independence, suspicion of outsiders, private property rights, local rule, states' rights, and distrust of government at all levels, especially the federal government. They see large carnivores as threats to their livelihoods and carnivore management as an intrusion of government into their lives.

In contrast with the Old West localists are the New West *environmentalists,* mainly people who have come to the region in the past 30 years with a more positive view of wildlife, especially carnivores.[9] People have been moving to the area in large numbers to live in beautiful settings away from congestion elsewhere. Many of these newcomers have settled in Teton County, though some reside in Sublette and other western Wyoming counties. These New West people are often less directly dependent on the land for their livelihoods and are not culturally or historically connected to it in the way that many localists are. Nevertheless, they still tend to have a strong connection to the landscape and its wildlife for aesthetic, recreational, and ethical reasons. Even those whose views do not fully align with this perspective share more with the environmentalist viewpoint than with the Old West localists.

The environmentalist group includes individuals and representatives of local, statewide, regional, and national conservation organizations and those who share the environmentally oriented views of these organizations. Although a clear minority, environmentalists reside in all four counties, with Teton County containing perhaps the greatest number of individuals with this viewpoint. People aligned with this group often value open space, animals, wildfire, wild vistas, undammed rivers, and nature conservation over exploitive uses of natural resources. They seek ecologistic values (emphasizing the biophysical patterns, structure, and functions of nature). Environmentalists show a different, more global kind of concern for nature and wildlife than that of localists. They also often have a more inclusive, universal attitude toward policy making, feeling that wildlife management should be accountable to a national rather than a local constituency.

Agency personnel, whose views represent a third main perspective in large carnivore management, include diverse people who live and work locally in district, state, regional, and national offices of wildlife, land, and resource management agencies. Although they often claim to be neutral and objective, adhering to the letter of the law in decision making and management, agency personnel have their own outlook. They are distinguished by a managerialist approach, that is, bureaucratic with a technical background. Their views about other aspects of the coexistence problem, however, may align with either the localist or the environmentalist camp or may fall somewhere in between. All four counties have resident agency personnel, mostly Wyoming Game and Fish Department (WGFD) and U.S. Forest Service (USFS) people, although other agency personnel are present as well. Their identifications and expectations vary, of course. State agency personnel tend to be in the localist camp, whereas federal agency people tend to lean toward environmentalist views, but not always. These distinctions are not always evident or clear.

These three participant groups are often in conflict and this is reflected geographically among the four counties. For example, Teton County stands in contrast to the other three counties in that many more environmentalists live in Teton County. Some participants feel that the region's resources should be exploited quickly, whereas others believe that they should be preserved and protected at any cost. Other people stand somewhere between these two extremes.

Perspectives as a Function of Education

The region's average educational levels vary.[10] According to 1990 data from the state and federal governments, education levels in western Wyoming are on par with those for the state and nation. This means that about half the residents of Sublette, Fremont, and Lincoln counties have a high school diploma. Teton County stands out in that it has a higher than average number of residents with college and graduate degrees, substantially more than any of the other three counties. According to studies by Jonathan Schechter of Jackson,[11] public school enrollment in Wyoming is declining in all but two of the state's 23 counties; in nine counties it has declined by more than 10 percent since 1991. In Teton County public school enrollment has increased by 16 percent, and in Johnson County it grew by 1 percent (eight students) in the same period.

Culture

The culture of Wyoming is unique in many ways, as discussed in the recent writings of journalist Samuel Western.[12] Wyoming's "Old West" mythology, captured in popular stories today, tells about the harshness of the land and the toughness of the mountain men, settlers, cowboys, and early entrepreneurs, the people who "created" the West. The myth includes accounts of large carnivores and how to deal with the threats they pose. Wyoming's early residents set the mold for how things should be understood and done in the West, and the state's culture strongly adheres to this formula even today.

This culture demonstrates its hold on people through their practices —what they say and do. The current culture still operates according to the formula of the Old West myth, which developed more than a century ago. These practices are organized into clearly visible patterns in communities and in people's lives, and they can be (and have been) studied and described. These patterns exist in the identities, expectations, and demands of citizens and are clearly evident in the culture's more permanent social, governmental, and political institutions, which embody the culture and give it structure and stability. They include the claims people make in public, in the media, and in the courts, and their use of symbols, including lore, stories, customs, western dress, and even the design of the state license plate.

Samuel Western described most Wyoming people as showing deeply rooted views of the type that we label localist, including strong protectionist policies for the state's beliefs, economy, and culture. One rancher in Western's account said, "Wyoming wants to get ahead but they want to stay back one hundred years, the way it seems. They don't want change." Western went on to say that Wyoming struggles with the modern reality that people, not natural resources, bring it wealth. Yet the culture is committed to a vigorous agricultural and resource extraction mindset. Wyoming's culture seems to be convinced that one more oil or gas well, one more dam, one more logging sale, or one more ranch or wheat field will bring prosperity. This belief is codified in a way of life, a mythology that, according to Western, has turned into a rigid, out-of-date ideology. In his view this unrealistic ideology has "stripped Wyoming of life-giving vigor." The author goes on to say that this ideology is out of touch with the facts of our time. It is a prison that keeps Wyoming from addressing many important matters, including change, making it difficult for the state to confront its cowboy past in light of its rapidly changing present.

A major part of this ideology is a profound hostility toward the federal government. Wyoming newspapers are full of articles about how bad the federal government is, how the "feds" need to be taught a lesson, or even thrown out of Wyoming. Localists often feel that federal agencies are not responsive to local needs and requests. They believe that federal actions violate their rules, their mythology, and their codes about "how things should be done." For example, federal managers accommodate wolves and grizzly bears, the very species that localists' forefathers worked so hard to eliminate. In fact, much of the local lore is about early settlers and cowboys overcoming the "cruelty and marauding ways" of mountain lions, wolves, and grizzlies.[13] Many locals are upset about the federal government "intruding" in their lives and "dictating" land use policies to them.[14] As a consequence, most issues about natural resource management in Wyoming are scripted in a state versus feds drama, a stereotype that is played out daily in innumerable ways. The script characterizes the locals as the good guys (cowboys) and the feds as the bad guys (rustlers).

A clear example is illustrated in a recent article in *Range.*[15] The article describes the USFS's intent to implement an order about how food should be stored in the backcountry of Bridger-Teton National Forest, and it includes county officials' summations of the feelings of many of their localist constituents. Sublette County commissioner Bill Cramer said that a "lot of what happens with predator conservation isn't driven by a concern for the species so much as a desire to eliminate other users of the national forests" and that "those who hold this view are arrogant and intrusive and have no regard for historic uses of the West that have occurred for generations." He claimed that residents of western Wyoming are "modern Indians" suffering government persecution, a feeling shared by many in the area. Expressing feelings of "victimization" and frustration, Stan Cooper, a Lincoln County commissioner, claimed that "you [i.e., outsiders and government] can't keep destroying our economic structure" through increased carnivore and land management regulations. Localists feel that since they are most affected by carnivores, they should have greater influence in their management. These statements articulate localists' beliefs that they are often not included in the decision process and that federal managers and agencies ignore their interests and concerns.

Furthermore, there is a widely held belief that Wyoming could "prosper if the federal government only let us alone," according to Western. Wyoming has a love-hate relationship with the federal government. On the one hand, Wyoming blames Washington, D.C., for everything it sees as wrong, and on the other hand, it seeks maximum financial help from Washington. Western argued that the longer Wyoming holds to its current belief structure, the "more embittered and protracted will be its battles with the federal government."

Another element in the ideology, said Western, is a belief that "agriculture remains a cornerstone in the state's economy." It does not seem to matter what economic and other trend data and statistics say about the facts of the matter (see, for example, chapter 6 of this volume). Wyoming wants to believe that agriculture is central to the future of the state. Western went on to say that "Wyoming's ideology requires presenting agriculture as the only means to achieve prosperity."

Overall, this ideology permits the state to remain in denial of the facts about the changing modern world and leads to a victim mindset, according to Western. The federal government or someone else—for example, environmentalists or outsiders—are the tyrants, the ones oppressing Wyoming, preventing the state from taking its rightful place in the world. This notion suggests that someone else controls the destiny of Wyoming, a stance that leads to a sense of resentment, hostility, and powerlessness. Not surprisingly, many young people leave Wyoming for a life elsewhere. As a result of these ideas and the self-confirming data selected to support them, Wyoming does not take responsibility for its policy and cultural future. The state sees itself as "Wyoming, the way the West was." Western concluded that Wyoming's culture today focuses on the wrong questions, thus inviting ineffective if not incorrect answers.

Another account of Wyoming's culture by Annie Proulx said that the state "is full of contradictions and anomalies. The state thinks small." "In its politics Wyoming is belligerently Republican, conservative—even reactionary—and fulminates against the heavy federal hand, people from somewhere else, taxes, and environmentalists." Finally, she said, "isolation is easy to mistake for independence."[16]

If we hope to develop effective policy for managing carnivores, we must acknowledge and understand the full range of perspectives, values, and subcultures of the residents, especially those of the dominant localist culture. Also, the social landscape of this region is changing rapidly, as it is throughout much of the American West, and we must take account of these trends as well. Developing policy that balances the diversity of these perspectives and values within a complex cultural milieu—with the goal of sustaining large carnivores and human communities—is the challenge that must be met by management agencies and citizens alike. Not all participants have direct influence over policy making, yet those who have an interest in how carnivores are managed may attempt to affect public decisions through political pressure, speaking at public meetings, writing letters to politicians, agencies, or newspapers, and contributing to organizations that work to influence policy.

Wildlife and Large Carnivores

Presently, some people feel that wildlife deserves special protection and restoration, but others disagree. Differences of opinions about the status, importance, and "meaning" of wildlife, especially large carnivores, are a significant feature of the human community in this region.

The Region's Wildlife

The region is highly prized by visitors and residents alike for its diversity of wildlife.[17] Many vertebrates are native, including 178 species of birds, 49 species of mammals, and numerous species of fish, amphibians, and reptiles. Innumerable invertebrates and plants are also represented. The region is unique in that extensive federal, state, and private lands comprise vast areas of contiguous, relatively high quality, wildlife habitat. This is one of the few places in the United States where there is still enough undeveloped and unfragmented habitat to support large populations of wild animals within a largely intact ecosystem. For example, according to the Wyoming Cooperative Fish and Wildlife Research Unit at the University of Wyoming, "Western Wyoming is home to the largest, most diverse ungulate populations in the western states."[18] This considerable prey base and the integrity of the region's natural landscape also make this one of the few areas in the United States that can still support large carnivores.

Large Carnivores

The major large carnivores in this area are mountain lions, wolves, and grizzly bears, the largest members of the cat, dog, and bear families in the region. Management of these three species is a high profile issue and the subject of heated debate. Other large- and medium-sized carnivores exist in the region as well, including black bears, lynx and bobcats, wolverines and otters, and coyotes and foxes.

Mountain lions (*Felis concolor*), also known as cougars or pumas, have always been relatively abundant and widespread. These large cats are solitary, territorial animals. Elusive by nature, they actively avoid people and livestock. They are very efficient predators whose diet mainly consists of ungulates. Mountain lions have always been hunted, but unlike wolves and grizzlies, were never eliminated from the region.

WGFD administers a mountain lion hunting season, takes responsibility for population monitoring, and controls individual animals that are thought to threaten people or livestock (see chapter 3). Management of mountain lions has received increased interest in recent years. For example, in spring 1999 a female lion and three kittens took up residence on Miller Butte just east of Jackson. Viewed by thousands, the Miller Butte mountain lions became big news. With unfortunately bad timing, however, WGFD increased the lion hunting quota that same year. This sparked outrage from environmentalists and brought into question the process and data used by WGFD to set hunting quotas. Another recent concern is that some residents feed deer, possibly attracting mountain lions to residential areas, habituating them, and endangering humans.[19]

Wolves (*Canis lupus*) are one of the most social carnivores. They were once distributed across North America. Demonized by the culture of the Old West, they were eradicated more than 60 years ago from Greater Yellowstone by bounties, trapping, poisoning, and shooting (see chapter 5). After much public discussion, wolves were reintroduced in 1995 into Yellowstone National Park as an "experimental" population under the U.S. Endangered Species Act. Since then, the wolves have thrived, primarily feeding on ungulates. Now this hardy and adaptable carnivore is recolonizing former habitat to the south of Yellowstone Park. In doing so, some wolves have preyed on livestock, and in these cases lethal control is typically the preferred management tool. Most of the area's residents did not want wolves reestablished and feel that the reintroduction was just another action that the federal government forced upon them. They often use the wolf as a political scapegoat to protest actions by a federal government with which they do not agree. Clearly, wolves carry a heavy symbolic and political burden.[20]

The success of the introduced wolves led the U.S. Fish and Wildlife Service (USFWS) to downlist the species in 2003 from endangered to threatened status, and the agency is now considering full delisting, which would remove protection under the Endangered Species Act. But the federal government and the state have failed to agree on a process for managing wolves after delisting. Wyoming has resisted assuming management authority without substantial funding from the federal government, arguing that it did not ask to have wolves reintroduced and should not be burdened with the expense of managing them. Also, in its proposed wolf management plan, the state classified wolves as predators outside national parks and wilderness areas, which would make them targets for unlimited killing. The USFWS rejected this plan in January 2004. Until such differences can be resolved, the wolves continue to be managed under federal jurisdiction.

Grizzly bears (*Ursus arctos*) are omnivorous animals, subsisting on a diet of insects, berries, nuts, and meat, primarily wild ungulates (see chapter 4). Formerly, they ranged throughout western Wyoming, but were hunted to near extinction outside the protection of Yellowstone National Park. Grizzlies have significantly expanded their range in recent decades, moving south as they will continue to do until stopped. Humans are the only substantial threat to the grizzly, and human-caused mortality is the main cause of adult grizzly deaths.[21]

The grizzly bear is currently a federally protected threatened species under the Endangered Species Act. Immediately after the bear was listed as such in 1975, Wyoming began efforts to have the species delisted. Delisting efforts are quite active today, and many federal managers feel the time has come to delist. However, environmentalists oppose delisting. If and when the bear is delisted, WGFD will assume management responsibility.

The Institutional System of Wildlife Management

Mountain lions, wolves, and grizzly bears are managed through the institutional system of wildlife management. One reason that people establish institutions is to carry out complex, cooperative tasks that require high levels of organization over time. Managing wildlife is one such task. As a result, an institutional arrangement has been built up around wildlife that sets out a stable, more or less orderly way in which species will be managed. The term "management" is somewhat of a conceit. Management does not act directly on animals. Instead, it controls, mandates, directs, or guides the behavior and actions of humans, which in turn have consequences for animals. Wildlife management is about managing ourselves and choosing what we will do to or for wildlife. Wildlife management institutions provide answers to two key questions: "How are we going to treat wildlife?" and "Who gets to decide?" Chapter 7 looks at this institutional system in detail, well beyond what is presented below.

Values

The practice of wildlife management is a human activity, an institutional arrangement created and used by people to achieve their values. This institutional system seeks answers to the two basic questions posed above, how wildlife will be treated and who gets to decide. The "treatment" of wildlife appears at first glance to be a technical question, but it is really about values, about who will benefit most from management, monetarily or otherwise. "Who decides" is also clearly a question of values, and first and foremost about power.

The institutional system largely controls how people interact with one another as they pursue their values by doing things to and for wildlife. The concept of an *institution* is examined in depth in chapter 7, but for the purposes of our discussion here it can be understood simply as a stable arrangement in the shaping and sharing of values among people. Note that this includes any group or collection of individuals that is organized for a central function or purpose, such as fulfilling the will of members or supporters or enforcing and implementing a government policy, mandate, or statute. Because institutions are human constructs, they directly reflect the perspectives and values of those who have the most influence over them, such as elites in the agencies, in politics, and in business.[22] The behavior of the institutional system of wildlife management reflects this fact of life and the competition for control among localists, environmentalists, and agency personnel. "Winners" get to answer the two questions in ways that serve their values and interests. They get to decide what people do, what happens to animals, and the nature of the institutional system.

Key organizational participants in the wildlife management system are introduced below. This is not a complete list or accounting, but these organizations are principals in carnivore management and are ultimately responsible for the creation and implementation of management policies. Their behavior also determines much of the behavior of the institutional system as a whole.

Wyoming Game and Fish Department

WGFD holds the dominant position regarding wildlife in the state. Actions by WGFD have significant impact on carnivore management. WGFD provides "a system of control, propagation, management, protection, and regulation of all wildlife in Wyoming."[23] WGFD manages all the state's wildlife species that are not federally protected, and it often participates in the management of those species that are federally protected. For example, WGFD manages mountain lions and shares management of grizzlies with the federal government, but wolves are currently managed by the USFWS (as required by the federal Endangered Species Act). However, even the Endangered Species Act is administered in close cooperation with the states. WGFD may become the primary manager for wolves and grizzlies in the future if current efforts to delist are successful. At present the department receives about two-thirds of its total earned revenues through the sale of hunting licenses and thus sees hunters as its main constituency. This view is strongly reflected in WGFD's decision making, public relations, and management. The department's pro-hunting stance is especially evident in its approach to carnivore management. For the most part, WGFD is more closely allied to the Old West, localist values, so this view of wildlife management is the current paradigm under which the state manages all wildlife.

The Wyoming Game and Fish Commission, a group of seven citizen representatives (lay people), oversees the WGFD and is responsible for the department's policy direction. Although no more than four representatives can be from one political party, most have close ties to livestock or hunting interests. The commission has the final vote on state management policy for carnivores. Since the governor appoints the commission, it often has a political agenda. The commission typically operates, as does the governor, with a strong localist and states' rights (i.e., antifederal) ideology. This in turn sets the tone for the department's interactions with other participants, including environmentalists, the public, and especially the federal government. Those with a states' rights viewpoint feel that, in short, the state should have all the authority and control over its wildlife (among other things). In practice this leads to turf wars between Wyoming, the national public, and federal agencies over carnivore management and endangered species restoration.

State management of wildlife is based on the above philosophical framework and set of practices that originated in the early 1900s and focused on game animals (i.e., those for which hunting licenses are sold). Today WGFD has expanded its interests to include many other species, but the historic framework continues to dominate and drive management practices. This philosophy and practice are deeply ingrained in many western state agencies, although the approach has been modernized somewhat with the aid of scientific methodologies, professionalization of staff, technologies such as computers, and bureaucratization. The actions of the commission and the department show that they view carnivores as a threat to ungulates and therefore a menace to hunters and to income from license sales. Additionally, WGFD sees carnivores as species to be hunted either as predators or trophy animals and managed lethally. If managed as trophy animals, large carnivores represent another source of income.

WGFD also follows managerialist or bureaucratic models of operation that enforce top-down views of problems and solutions. WGFD apparently believes that its experts alone are qualified to render professional judgments about wildlife management in the state. The combination of managerialism, top-down management, and the WGFD culture of expertise gives the agency an autocratic image and a distant, out-of-touch relationship with the public. In practice, its approach seems to be that the public should be "informed, consulted, or educated" as needed on a case-by-case basis. This formula results in an agency that does not always work productively with allies, much less opponents, to clarify and secure common interests in managing Wyoming's wildlife heritage.

U.S. Fish and Wildlife Service

The USFWS, an agency within the Department of the Interior, is charged with managing the nation's wildlife—including the "key objective [of] protecting endangered and threatened species and restoring them to a secure status in the wild."[24] This agency works in coordination with all other federal interests and the states to study and manage the wildlife for which it is responsible. The federal government, directly or indirectly, manages natural resources (including wildlife) in the West on USFS lands, Bureau of Land Management lands, and other federal lands. In Wyoming, the USFWS is currently responsible for managing wolves and grizzly bears, because they are federally protected species under the Endangered Species Act. The USFWS will oversee any transfer of management of these animals to the state if and when they are delisted.

The USFWS operates under a "federalism" philosophy, which is part of a long-term trend in centralizing decision making in the national government, a trend that was given boosts following the Civil War and World War II.[25] The rise of the modern environmental movement across the United States helped to bolster federalism. However, this trend is now shifting toward greater state and local management. Currently, the federal government and its agencies are undergoing a reinvention in their relationship to the states. It is important to note that there are actually very few USFWS personnel in western Wyoming.

U.S. Forest Service

The USFS manages the national forests owned by the federal government.[26] In western Wyoming, it manages Bridger-Teton, Targhee-Caribou, and Shoshone National Forests (figure 1.1). Part of the Department of Agriculture, the USFS operates under national legislation. One piece of guiding legislation, the Multiple Use Sustained Yield Act of 1960, requires that the USFS "develop and administer the renewable surface resources of the national forests for multiple use and sustained yield of the various products and services obtained from these areas."[27] In addition, the National Forest Management Act of 1976 mandates that the USFS "maintain viable populations of existing native and desired non-native vertebrate species in the planning area" (a viable population is defined as "one which has the estimated numbers and distribution of reproductive individuals to insure its continued existence is well distributed in the planning area").[28] These two mandates, along with other legislative directives, force the USFS to find some balance among many often conflicting goals. The agency has a relatively large presence and footprint in western Wyoming; it is clearly the most visible agent of the federal government in this region.

Other Federal Agencies

The Bureau of Land Management is the second most visible federal agency in western Wyoming. This organization, a multiple use land management agency, manages the sagebrush grasslands in the basins between the forested mountains (figure 1.1). It manages large blocks of land in Sublette and other counties, much of which is leased for sheep and cattle grazing. Often it is these livestock animals and those on adjacent USFS allotments that are attacked by large carnivores. Currently, the Bureau of Land Management is trying to manage the impacts of extensive oil and gas development on the region's mule deer, pronghorns, elk, and many other species.

Another federal agency with a significant presence in the region is the National Park Service, which manages Yellowstone and Grand Teton National Parks and the John D. Rockefeller National Parkway, which lies between the two parks. This agency is largely oriented toward protection, but at the same time exists to facilitate tourism and recreation. This dual mission causes the agency much consternation.

State and Local Governments

Other state agencies figure into carnivore management, including the Wyoming Department of Transportation (WYDOT), the Office of State Lands and Investments, which administers programs pertaining to resource management, economic development, and quality of life, and county governments. Roads are notorious corridors of death for wildlife and are especially disruptive to carnivore populations. Local governmental entities—such as the county commissions of Sublette, Fremont, Lincoln, and Teton Counties—play a major role in carnivore management. Town councils also play a key role. All regulate and manage the development of private properties, many of which border federally managed lands. Since these governmental bodies represent their communities and thus know what issues their constituents care about, they can act both as guides for their communities and as voices for the communities' concerns. They also attempt to influence large-scale management decisions through meetings with federal and state officials at which they communicate the wishes of their communities.

Nongovernmental Organizations

A broad spectrum of nonprofit and private groups actively work to influence policy making at all levels. Included are organizations that promote ranching; outfitting and hunting; recreational uses of public lands; environmental, wilderness, and wildlife conservation and protection; education; and other types of resource uses. These groups range in size and scope from local land trusts and small cattlemen's associations to larger organizations such as the Sierra Club and the Farm Bureau. They represent diverse values, perspectives, and interests and serve local, state, national, and sometimes international constituencies. They may act alone or as part of a coalition to influence policy. Because nongovernmental organizations often represent special interests, they have varying goals and expectations about outcomes.

Associations

Businesses, hunting clubs, and similar groups are associations active in large carnivore management policy. These often informal groups can have great influence under certain circumstances. Chambers of Commerce and petroleum, livestock, and outfitter associations are all key players too. The actual roles of all of these many associations are not well known. A complete listing and description would be helpful, given their potential influence.

Interactions among the Players

In conclusion, numerous organizations, associations, and government agencies influence and decide carnivore management. These many entities function to create a dynamic that is only partly visible. They operate in an arena that is highly fragmented in terms of authority and control. At least four federal agencies (USFWS, USFS, Bureau of Land Management, and National Park Service) and five state entities (WGFD, WYDOT, Office of State Lands and Investments, county commissioners, governor) are directly involved in conservation or land management activities, according to the Wyoming Conservation Directory published by WGFD in 2001.[29] The directory also identifies more than 40 nongovernmental conservation organizations that have a strong interest in how the area's resources are managed. This tally does not include the many agricultural and recreational associations, national conservation groups, and others that vie for a voice in management decisions. The governor's office and congressional delegates also often have tremendous influence over the development and direction of carnivore policy.

This broad spectrum of participants has an equally broad range of perspectives, roles in carnivore management, and relations to other participants. Each participant has distinct interests, biases, allegiances, and organizational culture, which often conflict with one another, giving rise to problems that at times seem insurmountable. The depth of passion, vehemence, and rigidity of positions raises the question of why people get so involved emotionally and vocally about carnivore management. As Harold Lasswell, a preeminent social scientist, noted, "Whenever there is a striking lack of proportion between an act and the reasons alleged for it, there is a presumption that some unconscious impulses are involved in the act."[30] There is clearly more involved in large carnivore management than first meets the eye.

Problems in Large Carnivore Management

Three trends are particularly important in the current context—changes in human use of the region, changes in wildlife populations and habitat, and changes in the institutional system of wildlife management. The combined effect of these trends and the conditions behind them suggests a future that is both more problematic and, potentially, more open to finding common ground. Currently, people and the institutional arrangements of wildlife management are struggling to address onrushing changes. Understanding these changes will be a first step in finding common interest solutions.

Changes in Human Uses of the Region

Humans have lived in Wyoming for thousands of years, but settlement of the western part of the state has increased substantially over the last 150 years, especially in the last three decades.[31] The original residents, descendants of peoples who moved south from Alaska after crossing from Asia, arrived more than 10,000 years prior to the expansion of settlers from both the east and west coasts in the mid- to late 1800s. The influx of colonists eager to exploit the mineral wealth, range lands, and forests began in earnest following the Civil War. During this time the region experienced increasing exploitation of its natural resources, a continuing trend (especially for oil and gas in recent years) that is bringing more people into conflict with one another and has implications for how large carnivores will be managed—eliminated, reduced, or allowed to expand.[32]

Settlement

The history of western Wyoming is fascinating. Early white visitors were trappers and explorers such as Jim Bridger, who began to pass through the area after the 1824 "rediscovery" of the Continental Divide crossing at South Pass by Jedediah Smith. Fur trading, one of the only ways to make a living in this area back then, became unprofitable by the 1830s, thus ending the era of the mountain men. The Oregon-California-Utah Trail was developed in the 1840s, as settlers made their way west. Settlement and organization of the territory increased greatly after the railroad came to the state in 1869.[33]

Statehood and States' Rights

The closing of the frontier was officially declared in 1890, the year Wyoming became a state. By then the majority of Native Americans had been moved onto reservations, increasing security and land availability for settlers, who flocked to the area. Livestock ranching became a dominant industry in the last decades of the century, as did logging, mining, and other forms of resource extraction.

In the four decades after Wyoming achieved statehood, the federal government took a substantial role in the organization and allocation of land. In addition to Yellowstone—the first national park, established in 1872—more federal lands were reserved in the national interest, including the Shoshone, Bridger, and Teton National Forests, and the National Elk Refuge in 1912. Grand Teton National Park was added in 1929 and expanded in 1949. The counties of western Wyoming were established during these early years, and resource extraction and agriculture dominated the culture and economy.

During this time most wildlife management at the state level focused on ungulates and other species that were hunted for recreation and food. The WGFD specialized in game management. This agency came into existence approximately 100 years ago. Large carnivores, seen as a threat to livestock and game species, were hunted exhaustively. The last wolf in Wyoming was killed in 1944, and grizzly bears were effectively eliminated outside Yellowstone National Park, where they had the protection of the 1894 Lacey Act.

The historically rooted states' rights ideology has strongly dominated all aspects of Wyoming's relations with other management participants, whether federal or not, up to the present.[34] This ideology, which emphasizes domain and control, has been vigorously asserted and prosecuted by the state government and its agents since statehood.

Federalism

In the two decades following the 1960 enactment of the U.S. Multiple Use Sustained Yield Act, a new era of management and conservation policy began in the American West. This law and others that followed through the early 1970s (e.g., the 1973 Endangered Species Act) greatly influenced the modern uses of federal lands. Although traditional activities such as logging and livestock rearing continued to dominate the economy, other land uses such as extraction of petroleum products played an increasingly significant role. At the same time there was a slow shift toward preservation and conservation of natural resources. A new national ethic calling for increased conservation began to make its way into policies and actions of public land managers in the West. This conflicted with traditional philosophies and land uses. As noted earlier, this conflict is still being played out today through issues such as large carnivore management.

Federal management actions that reflect national conservation values, such as the wolf and grizzly bear restoration actions of the 1980s–1990s, continue to change the way local residents use the region. Reintroduction of the wolf and federal protection of the grizzly bear have been very controversial.[35] Decision-making processes currently underway at the federal and state levels will determine much about the future management of carnivores in the region. Delisting both wolves and grizzly bears, now under consideration, and turning management responsibility over to the state will produce unknown consequences. The delisting process is proving to be the latest flashpoint in the seemingly endless conflicts over resource use that have become the norm in Wyoming. The conflict usually pits localists and WGFD against the federal agencies, although it is sometimes localists against the state as well, depending on the issue.

Communities

Throughout Wyoming's history, human cultures, perspectives, and values have changed dramatically. The social and economic structure of the western part of the state is still changing. The recent influx of new residents, many of whom are second homeowners and part-time residents, has led to a less homogeneous culture.[36] Many of the region's communities are highly polarized, with much posturing and grandstanding on all sides.[37]

Local culture is still dominated by the idea of rugged individualism, in which each person feels entitled to act as he or she wishes without having to adjust to outside changes. Allied to this is strong support of private property rights and states' rights, which many feel are being threatened or abridged by federal legislation that protects public land and wildlife. Many locals distrust the federal government and resent what they see as efforts to impose outside values on them. Another concern is that traditional users of the land feel disrespected and marginalized by current decision processes. Several people we interviewed expressed the view that outsiders, especially those from the eastern United States, "have no right" to tell them what decisions to make about resource use, even on federal lands. Much resistance to carnivore conservation in this area also stems from the fear that livelihoods and local control are being undermined.[38]

For some, recent change translates into a loss of social capital, real and perceived. Feeling that their way of life and security are threatened, people are increasingly responding to the uncertainty that accompanies change—specifically, the restoration of large carnivores—with full "heels-in-the-dirt" resistance.

Today

This stubborn dynamic and volatile history has resulted in distinctive economic and land use patterns. Western Wyoming is a rural place where localists have traditionally relied on natural resources to make a living, but this has shifted in recent years, according to the Wyoming State Economic Analysis Department. Employment data from the U.S. Bureau of Statistics show that between 1970 and 2000, employment on farms and in the agricultural sector overall—forestry, fishing, and mining—remained nearly level, while other sectors such as construction, services, retail, and government at least doubled in that period. The region's economy is coming to depend less and less on agricultural livelihoods.[39]

With the national parks and other vacation opportunities drawing millions of visitors each year, recreation and tourism are ever more important to local economies. In the last 30 years, the service economy—much of which caters to tourism and recreation—has grown substantially, outpacing all other sectors.[40] Nonearned sources of income are also becoming more important in Wyoming. Because they are not tied to local job markets, these sources offer some protection against the region's boom-and-bust cycles.[41]

Numbers of residents and visitors have expanded greatly in recent decades, as has their influence on local economies. As more people have moved to Jackson and other desired areas, property values have increased. Many towns have been feeling growing pains. As one Pinedale resident quipped in spring 2002, "The billionaires are pushing the millionaires out of Jackson and down to Pinedale." Such basic human perspectives must be recognized and understood contextually when looking at the social fabric of the region and identifying reasons for resistance to large carnivore conservation efforts.

Changes in Habitats and Wildlife

The preceding trends in human occupancy and land use are intensifying, and this is having profound impacts on wildlife and their habitats.[42] Native habitats are being degraded on larger scales than ever before. This causes declines in wildlife populations, dramatically so in some areas. Much of the region is still relatively "wild" land, which explains why large carnivores can still live there, but many ecological features are directly influenced by human decisions and activities. These activities are expected to have an even greater negative effect on wildlife in the near future.[43]

Lands

Land use is changing throughout the region.[44] Negative trends in land use include fragmentation of private lands and escalating use of public lands. Both have direct and indirect impacts on carnivores and other wildlife. Local ranchers have left the state with a "priceless gift of open space, a legacy with profound ramifications on the state's economy and character."[45] However, economic pressures make it difficult for already marginal ranching or agricultural operations to continue, causing ranch land to be sold, split up, and developed as subdivisions and ranchettes. Such development has social and economic effects on local communities and serious implications for wildlife. Increased housing construction and development at the public-private land interface also heighten the likelihood of large carnivores coming into contact—and conflict—with people.[46] Landscape fragmentation, development, and split ownership (surface vs. subsurface) ultimately diminish available wildlife habitat. The complex matrix of land ownership (figure 1.1) makes large carnivore management even more difficult.

A related problem is the increasing recreational use of public lands.[47] As more people travel the backcountry, it is not only more likely that they will meet up with large carnivores, but also that they will disturb or displace them. Associated trends, such as more roads and other infrastructure (e.g., power lines) have additional negative impacts on wildlife and habitats.

Elk and Mule Deer

Trends in management of elk are also leading to increased conflict. There is no doubt that large carnivores, particularly mountain lions, benefit from the abundance of prey, especially mule deer, in the region. The "artificially" high elk populations maintained by WGFD on the National Elk Refuge and on the winter feed grounds that they operate for elk provide an abundant prey base for wolves. The high ungulate populations facilitate continued expansion of large carnivore ranges to the south and east. Wolves have already visited several feed grounds, causing problems from the WGFD's point of view. Conflicts arise as both wolves and grizzly bears move south from the Yellowstone area onto Bridger-Teton National Forest, Bureau of Land Management, and private lands and into proximity of the feed grounds and private livestock.

Livestock

Abundant livestock also serve as prey. Bridger-Teton National Forest permits grazing of numerous cattle and sheep on forest lands. Sheep are more abundant at the southern end of the forest, and cattle are more numerous on northern allotments. Conflict arises when carnivores kill livestock that are grazed on legal allotments leased within the national forest from the federal government. Ranchers, who have been used to operating without large carnivores in the region for decades, now resent this assault on their livelihood and husbandry. This is an affront to both their wealth and skill.

Increasing changes in the landscape, reduction of wildlife habitat, and growing carnivore impacts on livestock and wildlife promise to bring even more conflict in the future unless something is done. Today, outside of a very few individuals, no one among community leaders, agency officials, or carnivore managers seems to have adequate knowledge of these trends or practical knowledge about effective actions that can be taken. Cooperative methods for directing and managing community growth, as well as livestock management, to avoid evident problems are known and used elsewhere. The highly conflictual context of western Wyoming makes it difficult to put these methods into practice effectively, in large part because of the institutions used to manage carnivores, wildlife, and other community matters.

Changes in the Institutional System of Wildlife Management

The institutional system for managing wildlife has changed dramatically since early trappers, explorers, and settlers appeared on the scene more than 150 years ago. Today this institutional arrangement comprises many individual and organizational participants, each of which takes part in both separate and shared actions and "decision streams" that affect large carnivores as well as people. Each individual and organizational actor, whether federal, state, or local, has special interests and capacities that may or may not be compatible or complementary with those of others. When contradictory interests collide, the system does not function to find common ground.

Functioning

The structure and function of institutions determines what happens to wildlife in the region. The institutional participants that have primary responsibility over carnivore management are the USFS, USFWS, and Bureau of Land Management at the federal level and WGFD at the state level. All tend to be reactive rather than proactive in their problem solving, an approach that forces them into a nearly constant state of crisis management and almost ensures continued conflict. Longer term challenges are not well anticipated, so little is done to avoid, prevent, minimize, or mitigate them.

Problems

Problems that hinder performance are clearly evident within the agencies. First, limited "institutional memory" results in repeating past mistakes. Second, internal review processes, which could steer the agencies in new, more effective directions, seem to be minimal. Third, there is a lack of leadership, specifically, a lack of strategic policy vision. Fourth, existing leaders seem reluctant to support innovative employees. These problems cause frustration, low morale, and in the worst cases, the loss of valuable employees.[48]

A case in point: WGFD embodies the localist, states' rights culture and considers hunters and ranchers the department's only true constituents. Ideologically, WGFD resents and rejects federal jurisdiction over certain aspects of large carnivore management, yet it is very willing to accept federal money to carry out wildlife management programs. The department portrays itself as a victim, typically blaming monetary or manpower limitations for its shortfalls, while ignoring its own internal cultural, structural, and operational shortcomings. As one anonymous WGFD employee stated in March 2002, the department's view of scientific data is that "less is best" because more data might force the department to change and adapt.

Consequences

Overall, the institutional system shows muddled leadership, goal confusion, and weak implementation in the field. There is little evidence that corrective evaluation or systematic learning from experience is used to upgrade problem solving. There are reasons for this underperformance. As noted earlier, the arena is highly fragmented in terms of authority and control. The many decisions and actions involved take place at different social and political levels. This creates competition, lack of coordination (and even antagonism), and weak overall decision making and management. This basic problem is especially evident when the agencies try to implement controversial policies, such as the USFS's food storage order described earlier. These observations raise serious questions about how to allocate authority and control for wildlife management among agencies and how to create a framework —an institutional system—for their interaction that is efficient, effective, and equitable. At present, there is little recognition within the agencies that this institutional problem exists, much less an awareness or an active search for ways to improve and perform more effectively. Since this institutional system shapes present and future policy, it is incumbent on everyone to make it work well to serve common interests to the maximum extent practicable.

To summarize, the rapidly changing human and ecological landscape affects people and wildlife often in harmful ways. The current institutional system of wildlife management is basically unable to keep up with both the rate of change and demands for better performance. It is unable or unwilling actively to clarify and secure common interest solutions for managing carnivores. Improving management will require participating organizations to recognize these weaknesses and amend their approach to policy making.

Recommendations

Managing large carnivores at present is less a problem of animal biology or control and more a problem of cultural perception, decision process, and institutional dilemma. This becomes inescapably evident when the full context of management is appreciated and considered.

Before offering recommendations, a tentative problem definition (a refinement of the problem) is needed. To be sure, there are real conflicts between people and wolves, bears, and mountain lions, and we must find answers to the very real questions about how to control animals and under what circumstances large carnivores should be killed. Lethal actions make great headlines. However, stepping back from the headlines and the immediate conflict and taking a larger, contextual view can help us to focus our attention on other questions, such as, What do carnivores mean, factually and symbolically, to people? The current institutional system and its processes of decision making do not seem inclined to ask or answer such questions systematically or effectively. Both issues—the actual problems caused by carnivores and the meaning of carnivores to people—must be addressed if we are to live with these animals. Lingering, destructive conflict and mistrust indicate that the current management system is not as effective as it needs to be in understanding and addressing these matters. As noted in chapter 1, coexisting with large carnivores is problematic precisely because of people and their polarized perspectives on these questions. The basic problem is about how we interact with one another over troubling public issues and collectively decide how we want to live.

If this problem definition is accurate, then the next step is to invent, evaluate, and select solutions on the ground, the only place that really matters. This requires creativity. Solutions will mean changes in perceptions, institutions, and decision making. We believe that successful solution strategies can be developed given the context. We recommend that all participants (1) use reliable knowledge of the context to make decisions and carry out actions, (2) increase the basic structural capacity of wildlife management institutions, and thus (3) improve the quality of decision-making (management) processes.

Increasing and Using Contextual Knowledge

Carnivore managers, agency leaders, citizens, scientists, activists, ranchers, and others must ground their problem definitions and their search for solutions in the entire management context, not selective parts of it that give an advantage to their special interests. Trying to solve this complex problem with a limited map of the context—that is, the operational environment—is like trying to traverse Teton Wilderness with little knowledge of the landscape, animals, and dangers, and without food, trails, or compass. To the extent that we fail to create and use a reliable contextual map, we are "traveling blind." We all tend to undervalue the significance of knowledge we don't have, yet a good map can help everyone, and especially officials, avoid biases, favoritism, or selectivity in understanding the situation and in making decisions. Despite the obvious need to design policy that is contextually realistic, too often management decisions are made with little regard explicitly for human social factors—essential elements of the context. This failure to identify all of the relevant information seems to be caused by limited time, skills, and resources, policy preferences that ignore the context, and political and other biases that blind participants to important aspects of their operating environment. We all have biases, such as loyalties to our employers, professions, and friends. These too are part of the context and must be mapped and accounted for in decision making and management. This requires substantial self reflection.

In this chapter we have introduced these concepts and sketched out a preliminary contextual map that should be useful to all participants in large carnivore management in the region. The map must be refined over time and detailed on a case-by-case basis. To do this realistically requires data. The agencies, in particular, must broaden their fact-gathering activities about the context, including themselves as players in management. The results will give everyone more reliable and comprehensive information for decision making. It will help all parties to set practical goals and realistically identify points of leverage and potential pitfalls in their collective efforts to improve management. A better contextual understanding will enable managers and others to develop policy that effectively addresses problematic situational variables. For example, agencies can focus their efforts on situations that are already high in conflict or most likely to become contentious. This will help to reduce existing conflict and prevent additional conflict (real or perceived) from occurring. Overall, increasing contextual knowledge will provide all of us with a better operational map that can be used to negotiate the difficult terrain through which we are traveling.

Increasing Institutional Capacity

Institutions are the means through which people work to achieve their common goals. Institutional capacity must be upgraded on many levels in carnivore management. One reason that a reliable contextual map does not exist at present is that the people who most influence decision making either do not see the need for one or do not know how to build one in an inclusive, participatory way. We recommend, first, that leaders use their positions to encourage people to experiment and challenge themselves and their views in relation to carnivore management in a constructive manner. This means that leaders should skillfully lead respectful, genuine discussions and problem-solving activities. It is their responsibility to provide a supportive environment for learning and positive change as well as methods and incentives to do so. We must find better ways to incorporate new knowledge, ideas, and practices into management. Through active leadership, institutions can be transformed and made more effective. Chapter 7 expands on ways to improve the institutional system of wildlife management.

Second, the skills of individual managers could be strengthened and better used so that their actions will be problem-oriented and contextual in a real-world, practical way.[49] This may require some people to "unlearn" the less effective approaches they currently use. Presently in short supply are managers with the skills to understand and sympathize with, work with, and earn the respect of a broad range of constituents, who may hold different and conflicting views, without sacrificing their integrity or professional commitment to common interest goals. Civic-minded managers who assume "coordinating" or "bridge-building" roles can go a long way toward improving trust and cooperation as well as resolving legitimate people-carnivore conflicts. The special skills required by these roles can be taught, learned, and applied.

Third, institutions should be more involved in "prototyping," or what we might also call "practice-based" problem solving. This strategy, which should be inclusive and respectful of everyone involved, calls for informed invention, trying new practices, and learning as we go. Cooperative programs can be set up to address conflicts between carnivores and livestock, or carnivores and recreational forest users, or other on-the-ground problems. Better leadership, stronger worker skills, and prototyping will all allow the institutional system to be more responsive to the context and the needs of people living with and most affected by the presence of large carnivores. These "tools" are perhaps our best options. They can provide guidance and feedback that can be used to promote dialogue and understanding, and most importantly, improve actual management practices.

Improving Decision Making

The quality of the processes by which decisions are made greatly influences the likelihood of successful policies. To define and solve problems effectively, decision processes must function in a realistic and timely way. Management agencies need to reevaluate how they interact with the public and how they create and implement their policies. They will likely need outside help in performing such an evaluation. The best kind of evaluation is ongoing and independent. The results will likely reveal that more effective, more inclusive, and more cooperative problem-solving means are available to them than those they currently employ. An evaluation is also likely to show that better, more contextual approaches are possible and indeed desirable. Decision making must be initiated at the right time, informed by necessary and sufficient information, and discussed in open debate. Management decisions must be realistically formulated, adequately enforced and implemented, fairly evaluated, and terminated as appropriate. This entire process must have sufficient staff and resources.

We recommend using a tried-and-true way to improve decision making: find and build on existing models that people agree are successful. For example, the statewide Sage Grouse Working Group that was active from 1999 to 2002 was reported to fit many of these criteria.[50] This group was formed to help management agencies determine how to protect sage grouse while actively managing and, in some cases, developing the species' habitat. With representatives from agencies, the private sector, and the community, the team worked together to come up with management guidelines that would likely be effective and palatable to all the participants. This example may well be a case in which the common interest was clarified and eventually secured. Many other examples exist, according to Ron Brunner and colleagues, who recently looked at examples throughout the American West.[51]

We also recommend that decision making be improved by striving to meet well-known standards. That is, decision making must be open, fair, comprehensive, reliable, creative, rational, integrative, effective, constructive, timely, dependable, independent of special interests, fully contextual, respectful, balanced, prompt, ameliorative, reputable, and honest.[52] Overall, the process should aim for a reputation of honesty and public service. It is important for leaders to recognize that many individuals would be willing to engage in productive conversations and problem solving if they were approached respectfully by someone who is willing, in their estimation, to give them a fair hearing. A process that strives to meet these standards offers the best means to engage people in addressing common interest problems.

Lastly, we recommend that decision making should avoid common pitfalls. These include biased problem definitions, inadequate analysis of problems and limited search for solutions, poor coordination and overcontrol by government and powerful players, weak implementation, insensitivity to constructive criticism, and pressure to continue unsuccessful policy, to mention only a few.[53] High quality decision-making processes promote trust and the buildup of social capital.

Conclusion

Large carnivore management takes place in a complex and dynamic context. Many people in this context, including residents of western Wyoming as well as the national public, have made it very clear that they want large carnivores to be part of their environment. They want people and carnivores to coexist. The management problem is how to bring this goal into reality, given the context, which also includes people who reject and oppose that goal. As large carnivores disperse south from the Yellowstone region onto both public and private lands, carnivore conservation moves from an abstract vision to a very real, on-the-ground management challenge. Legitimate problems arise from living with carnivores. Large carnivores kill livestock and cause problems for human safety and well-being. As well, harmful conflict stems from incompatible, competing worldviews, values, and symbolic meanings. Finally, the unproductive conflict persists because of weak institutions and decision making. The present decision processes and the institutional system of wildlife management have failed to clarify or secure the common interest. Perpetuating these management practices and processes will not result in the resolution of people's differences nor in progress. Instead, it will likely cause more negative effects for both people and carnivores in the future.

Not only are improvements desirable, they are possible. Realistically understanding the context, mapping it in a timely, comprehensive fashion, and using this knowledge are essential to improve management and guide it toward balancing people and carnivores. This requires skillful, civic-minded leadership and citizen involvement in all aspects of the management process. To foster such a process, enhance the prospects for constructive dialogue, and build toward a successful outcome, individuals, groups, and the overall system of wildlife management must function better. Specifically, this requires increasing the availability of contextual information, upgrading institutional capacity, and bringing about better decision making. We cannot achieve sustainable human communities or viable carnivore populations, much less coexistence, if we choose to do nothing. The other choice—to do something about the problem—requires us to ground our decisions in realistic knowledge about the full context. This means improving information gathering, processing data, and disseminating it widely to all interests. This strategy must be integrated fully into management as part of an overall problem-oriented, contextual, multimethod effort to understand problems and find realistic, enduring solutions. Such an approach would be acceptable to a majority of citizens and result in long-term coexistence of people and large carnivores.

Our analysis and recommendations are meant to aid the search for common interest outcomes. It is through open dialogue and genuine problem solving that people's common interests can best be clarified and secured. Those responsible by law for ensuring the conservation of large carnivores—agency employees, administrators, and elected officials—must find better ways to manage these species and human interactions with them. They must build support across interest groups to find integrated, "win-win" outcomes.

References

1. Numerous social scientists have written on the importance of context and why it must be mapped and understood to improve natural resource management policy. The "principle of contextuality" is examined by T. W. Clark, 2002, *The policy process: A practical guide for natural resource professionals,* Yale University Press, New Haven, 29. The dimensions of context are often neglected or overlooked in management cases. See also G. Honadle, 1999, *How context matters: Linking environmental policy to people and place,* Kumarian Press, Bloomfield, CT; and T. W. Clark et al., 2000, "Interdisciplinary problem solving in carnivore conservation: An introduction," 223–240 in J. Gittleman et al., eds., *Carnivore conservation,* Cambridge University Press, Cambridge. This article explicitly calls for mapping the context in carnivore management cases. For a general overview of human relations to predators, see D. Quammen, 2003, *Monster of God: The man-eating predator in the jungles of history and the mind,* W. W. Norton, New York.

2. The practical and theoretical basis of context has been well-described in the literature. A guide on how to map context is given in Clark, *The policy process,* 29.

Examples of researchers and managers mapping context are in T. W. Clark, A. R. Willard, and C. M. Cromley, eds., 2000, *Foundations of natural resource policy and management,* Yale University Press, New Haven. The groundwork for this kind of mapping was developed in H. D. Lasswell, 1971, *A pre-view of policy sciences,* American Elsevier, New York; and H. D. Lasswell and M. S. McDougal, 1992, *Jurisprudence for a free society: Studies in law, science, and policy,* 2 vols., New Haven Press, New Haven.

3. For physical descriptions of the climate, topography, geology, and vegetation of the region, see D. Knight, 1994, *Mountains and plains, the ecology of Wyoming landscapes,* Yale University Press, New Haven. Temperatures were summarized from 1971–2000 climate data for Wyoming, Wyoming State Climatologist Office Web page, http://www.wrds.uwyo.edu/wrds/wsc/normals/monthly_normals.html, (accessed November 5, 2002). For additional summaries, see R. B. Keiter and M. S. Boyce, 1991, *The Greater Yellowstone Ecosystem: Redefining America's wilderness heritage,* Yale University Press, New Haven; T. W. Clark and S. C. Minta, 1994, *Greater Yellowstone's future: Prospects for ecosystem science, management, and policy,* Homestead Publishing, Moose, WY; and R. Reese, 1984, *Greater Yellowstone, the national park and adjacent wildlands,* Montana Geographic, Helena.

4. Quotations from Reese, *Greater Yellowstone, the national park and adjacent wildlands,* Keiter and Boyce, *The Greater Yellowstone Ecosystem: Redefining America's wilderness heritage,* and Knight, *Mountains and plains, the ecology of Wyoming landscapes,* as cited by Paul Schullery in The Greater Yellowstone Ecosystem, National Park Service, available at http://biology.usgs.gov/s+t/noframe/r114.htm#23507, (accessed November 29, 2004).

5. Charture Institute, 2003, *The Jackson Hole almanac: Facts and data about the Teton County area,* Charture Institute, Jackson, WY; R. Rasker and B. Alexander, 2003, "Getting ahead in Greater Yellowstone: Making the most of our competitive advantage," Sonoran Institute, Bozeman, MT; this document is available on the institute's Web site http://www.sonoran.org/programs/si_se_program_grt_yel.html, (accessed November 29, 2004).

6. See K. Byrd, 2002, "Mirrors and metaphors: Contemporary narratives of the wolf in Minnesota," *Ethics, Place, and Environment* 5(1), 50–65. For a guide to participants in conservation see National Wildlife Federation, 2003, *Conservation directory,* Washington, D.C.; and S. R. Brown, 2003, "Workshop on large carnivore conservation (summary report to participants), Department of Political Science, Kent State University, Kent, OH.

7. S.R. Brown, 2003, "Working on large carnivore conservation," summary report to participants, Department of Political Science, Kent State University, Kent, OH.

8. R. Slotkin, 1992, *Gunfighter nation: The myth of the frontier in twentieth century America,* Harper, New York.

9. S. R. Kellert et al., 1996, "Human culture and large carnivore conservation in North America," *Conservation Biology* 10, 977–990.

10. Information and commentary on educational achievement in the region is given by S. Western, 2002, *Pushed off the mountain, sold down the river,* Homestead Pub-

lishing, Moose, WY; and P. Krza, 2003, "Wyoming at a crossroads," *High Country News,* February 17, 1, 14–17. Statistical education data were summarized from Wyoming Department of Administration and Information, http://eadiv.state.wy.us/ and U.S. Census Bureau, http://www.census.gov/ (accessed November 29, 2004).

11. Charture Institute, 2003, *The Jackson Hole almanac: Facts and data about the Teton County area,* Charture Institute, Jackson, WY.

12. These publications describe the defining characteristics of the dominant culture and "Old West" mythology of the region: Western, *Pushed off the mountain, sold down the river;* D. Davy, 2002, *Cowboy culture: A saga of five centuries,* University of Kansas Press, Lawrence; P. N. Limerick, 1987, *The legacy of conquest: The unbroken past of the American West,* Norton, New York; R. Slotkin, 1973, *Regeneration through violence: The mythology of the American frontier, 1600–1860,* Wesleyan University Press, Middletown, CT.

13. T. W. Clark and D. Casey, 1992, *Tales of the grizzly: Thirty-nine stories of grizzly bear encounters in the wilderness,* Homestead Publishing, Moose, WY; D. Casey and T. W. Clark, 1995, *Tales of the wolf,* Homestead Publishing, Moose, WY.

14. This sentiment is felt by many local residents throughout the region and was expressed in both interviews and public meetings: Maury Jones, outfitter in Afton, pers. comm., March 2002; Bill Cramer, Freemont County Commissioner, and the testimony of outfitters at the Freemont County Commissioners' meeting, March 2002; Citizen comments at a public meeting about implementation of a backcountry food storage order on the Bridger-Teton National Forest in Afton, March 2002.

15. This conflict is outlined in C. Urbigkit, 2002, "The coup counties," *Range,* Fall, 30–31. Numerous other examples appear in the press. For a very recent example, see B. Hassrick, 2004, "Rancher: Wolf agent 'trespassed,'" *Cody Enterprise,* February 26.

16. A. Proulx, 2003, "Wyoming, the cowboy state," 495–508 in J. Leonard, ed., *These United States,* Thunder's Mouth Press, New York.

17. These publications provide excellent descriptions and discussions of the wildlife and habitats of western Wyoming and were relied on for the information in this section: P. Schullery, 1992, *The bears of Yellowstone,* High Plains Publishing Company, Worland, WY; WGFD, 1998, "Wyoming's threatened and endangered species on the edge," *Wyoming Wildlife,* June, 15–34; T. W. Clark et al., 1999, *Carnivores in ecosystems: The Yellowstone experience,* Yale University Press, New Haven; T. W. Clark and M. R. Stromberg, 1987, *Mammals in Wyoming,* University of Kansas, Lawrence; N. Matson, 2000, "Biodiversity and its management on the National Elk Refuge, Wyoming," 101–138 in T. W. Clark, D. Casey, and A. Halverson, eds., *Developing sustainable management policy for the National Elk Refuge, Wyoming,* Yale School of Forestry and Environmental Studies Bulletin No. 104.

18. H. Sawyer, S. Anderson, and F. Lindzey, 2001, "Sublette mule deer study, Wyoming Cooperative Fish and Wildlife Research Unit, Laramie; H. Sawyer, S. Anderson, and F. Lindzey, 2000a, Jackson Hole pronghorn study," Wyoming Cooperative Fish and Wildlife Research Unit, Laramie; for an online summary of these reports see http://uwadmnweb.uwyo.edu/fish_wild/report/completed_projects.html (accessed November 29, 2004).

19. The story of the Miller Butte mountain lions and the almost simultaneous increase in hunting quotas is documented in chapter 3. See also D. Darr, 2002, "Locals continue to fight to end wildlife feeding," *Jackson Hole Guide,* March 13, A19.

20. The symbolism of and human attitudes toward wolves, grizzly bears, and mountain lions are further analyzed by Kellert et al., "Human culture and large carnivore conservation in North America." Wolf management is politically contentious, as seen in A. M. Thuermer, 2002, "County legislators back wolf criticism," *Jackson Hole News,* March 20, A16; W. Royster, 2002, "Teton wolf comments not in step with state," *Jackson Hole News,* August 14, A28; D. Darr, 2002, "Public offers divergent views on state wolf plan," *Jackson Hole Guide,* August 14, A30; W. Royster, 2002, "Wolf manager dispels myth about slaughter," *Jackson Hole News,* August 14, A29.

21. Experts who have spent many years studying the grizzly have written of its biology in D. Mattson and M. Reid, 1991, "Conservation of the Yellowstone grizzly bear," *Conservation Biology* 5, 364–372; Schullery, *The bears of Yellowstone.* The recent expansion of grizzly range and resulting conflicts have been well-documented by local newspapers: "Grizzlies continue southern expansion outside GYE," 2002, *Jackson Hole Daily,* August 20, 6; "Grizzlies expanding range," 2002, *Jackson Hole Daily Guide,* August 20, A12; J. Wilhelm, 2002, "Bears on the porch!" *Pinedale Roundup,* September 5, 1, 19; J. Wilhelm, 2002, "Grizzly bear killed in Wyoming Range," *Pinedale Roundup,* August 22, 3. Improvements in the process of managing grizzlies are discussed by S. Primm, 1994, "Grizzly-livestock conflicts on Togwotee Pass, using research to find solutions," *NRCC News* (Northern Rockies Conservation Cooperative, Jackson, WY). There are many other scientific and government accounts in the literature.

22. The organizational dynamics and cultures of institutions are discussed by T. W. Clark, 1997, *Averting extinction: Reconstructing endangered species recovery,* Yale University Press, New Haven; J. N. Clarke and D. C. McCool, 1996, *Staking out the terrain: Power and performance among natural resource agencies,* State University of New York Press, Albany; A. Halverson, 2000, "The National Elk Refuge and the Jackson Hole Elk Herd: Management appraisal and recommendations," 23–52 in Clark et al., *Developing sustainable management policy for the National Elk Refuge, Wyoming.*

23. Much information about WGFD can be found on the department's Web site: http://gf.state.wy.us/ (accessed November 29, 2004); and in a book by N. Blair, 1987, *History of wildlife management in Wyoming,* WGFD, Cheyenne. Other information about WGFD was collected through personal communications with B. Holz (WGFD wildlife supervisor), M. Bruscino (WGFD Bear Management Officer), M. Gocke (WGFD public information specialist), March 2002. Budgetary information collected from WGFD, 2001, *Statement of revenue & expenditures for the periods ending December 31, 2001,* WGFD, Cheyenne; C. Frank, 1999, *WGFD financial status briefing statement,* WGFD, Cheyenne; L. L. Kruckenberg, 2000, *Special report–Game and Fish's financial situation in the year 2000 and beyond,* WGFD, Cheyenne.

24. Information about this agency can be found at the USFWS Web site, http://www.fws.gov/ (accessed November 29, 2004).

25. The history of environmental legislation is outlined by R. V. Percival et al., 2000, *Environmental regulation: Law, science, and policy,* 3rd ed., Aspen Law and Business, New York, 101–113. See also R.B. Keiter, 2003, *Keeping faith with nature: Ecosystems, democracy, and America's public lands,* Yale University Press, New Haven.

26. Information about this agency can be found at the USFS Web site, http://www.fs.fed.us/ (accessed November 29, 2004).

27. National Multiple Use Sustained Yield Act, Public Law 86-517, 86th Cong. (June 12, 1960).

28. National Forest Management Act, P.O. 94-588, 94th Cong. (October 22, 1976).

29. The directory is published to "further communication and networking among the state's conservation community." J. Lawson et al., 2001, *Wyoming conservation directory,* WGFD, Cheyenne.

30. H. D. Lasswell, 1977, "The triple appeal principle," *Harold D. Lasswell on political sociology,* D. Marvick, ed., University of Chicago Press, Chicago, 223.

31. A detailed account of Wyoming's history is provided in T. A. Larson, 1990, *History of Wyoming,* 2nd rev. ed., University of Nebraska Press, Lincoln.

32. Oil and gas concerns are reported by G. Wilkes, 2002, "Threat to deer migration bottleneck halts Trapper's Point oil and gas sale," *Pinedale Roundup,* August 22, 1; R. Huntington, 2002, "Oil, gas bids offered in wildlife corridor," *Jackson Hole Guide,* September 4, A3; R. Odell, 2001, "Forest supervisor faces down oil drilling," *High Country News,* March 26, 5; W. Royster, 2002, "Forest to draft plan for drilling in Bondurant," *Jackson Hole News,* March 27, A11.

33. See Larson, *History of Wyoming,* and also C. Cromley, 2000, "Historical elk migrations around Jackson Hole, Wyoming," 53–65 in Clark et al., *Developing sustainable management policy for the National Elk Refuge, Wyoming.*

34. Percival et al., *Environmental regulation: Law, science, and policy;* Clarke and McCool, *Staking out the terrain: Power and performance among natural resource agencies;* F. McDonald, 2002, *States' rights and the Union: Imperium in imperio, 1776–1876,* University of Kansas Press, Lawrence; Western, *Pushed off the mountain, sold down the river;* Wyoming Outdoor Council, 2003, "SF 97 Wyoming's wildlife," Frontline 2003 Legislative Report, Spring, 9; Wyoming Conservation Voters, 2003, *2003 Legislative scorecard,* Wyoming Conservation Voters, Casper, WY.

35. Numerous news articles have detailed the contentious nature of carnivore management in recent years. Those listed here are only a small sample of the debate: C. Urbigkit, 2002, "Officials outlaw grizzlies, wolves in Freemont County," *Casper Star Tribune,* March 8, A1; Urbigkit, "The coup counties"; J. Lawson, 1998, "Grizzly delisting charged with emotion," *Wyoming Wildlife News,* Sept.–Oct., 7; W. Royster, 2002, "State agency puts wolves in crosshairs," *Jackson Hole News,* September 18, A1, A30; A. Thuermer and W. Royster, 2002, "Enzi decries wolf," *Jackson Hole News,* July 17, A1, A18.

36. The increase in second-home ownership in the region has been highlighted in the press: G. Wilkes, 2002, "Sublette leads state in percent of 2nd homes," *Pinedale Roundup,* August 8, 3; and in published reports such as: Wyoming Open Spaces Research

Group, *Second home growth in Wyoming, 1990–2000,* April 2002, Wyoming Open Spaces Research Group, Publication B-1120.

37. As noted by Clark and Minta, *Greater Yellowstone's future,* 31, and supported by discussions with numerous agency, community, and organizational participants.

38. See endnote 12. This feeling is also evident in local and regional newspaper editorials and letters to the editor: R. O'Toole, 2002, Opinion: "Locals deserve more influence over forests," *Jackson Hole Guide,* March 13, A6; M. Jones, 2002, Letter to the editor: "Taken more than we can bear," *Casper Star Tribune,* March 17, E3.

39. See G. Wilkes, 2002, "Services grow while ranching declines in Sublette County," *Pinedale Roundup,* May 2, 15.

40. See W. E. Riebsame, H. Gosnell, and D. M. Theobald, eds., 1997, *Atlas of the New West: Portrait of a changing region,* Norton Press, New York.

41. See *The importance of non-earned income in Wyoming,* Economic Analysis Division, Department of Administration and Information, State of Wyoming, Cheyenne, Special report 1, August 17, 2001.

42. These trends and effects are discussed by A. J. Hansen et al., 2002, "Ecological causes and consequences of demographic change in the New West," *BioScience* 52(2), 151–162.

43. Elk and mule deer trends are summarized by Cromley, "Historical elk migrations around Jackson Hole, Wyoming." Newspaper accounts of carnivore range expansion (see endnote 21) and conflict with cattle and sheep help to track these trends: D. Darr, 2002, "Grizzly killed for livestock conflicts," *Jackson Hole Guide,* August 21, A3; Wilhelm, "Grizzly bear killed in Wyoming Range"; W. Royster, 2002, "Feds aim to let ranchers shoot wolves," *Jackson Hole Daily,* March 8, 1, 3; W. Royster, 2002, "Third wolf killed in Dunoir," *Jackson Hole Daily,* August 1, 3.

44. In a recent study by American Farmland Trust, Sublette and Fremont counties ranked in the top 25 of 263 western counties "at risk of losing ranchland to low density development." This concern is reported by R. Shaul, 2002, "Ranchlands lost to subdivision, Sublette ranks 13th in region for ranchlands at risk," *Pinedale Roundup,* July 2, 3. Other perspectives on land use change are found in V. K. Johnson, 2001, "Trends in rural residential development in the Greater Yellowstone Ecosystem since the listing of the grizzly bear, 1975–1988," *Yellowstone Science* 9(2), 2–9; Second Home Growth in Wyoming, 1990–2000; G. Wilkes, "Sublette leads state in percent of 2nd homes"; G. Wilkes, 2002, "Smaller lots new trend as county subdivides," *Pinedale Roundup,* June 13, 1.

45. Western, *Pushed off the mountain, sold down the river,* 97.

46. This problem was emphasized in personal communications with Mike Jimenez (USFWS), Timm Kaminski (USFS), and Kim Barber (USFS), March 2001. The concern that "as more people crowd into the west's wild places, conflicts with black bears and grizzlies increase," is explored in T. Christie, 2002, "Problem bears," *Wyoming Wildlife,* August, 10–17.

47. Susan Marsh, Bridger-Teton National Forest Recreation Staff Officer, pers. comm., July 2003; Johnson, "Trends in rural residential development in the Greater Yellowstone Ecosystem since the listing of the grizzly bear, 1975–1988."

48. The institutional system's features and problems are discussed by F. B. Samson and F. L. Knopf, 2001, "Archaic agencies, muddled missions, and conservation in the 21st century," *BioScience* 51(10), 869–873. Chapter 7 covers the institutional system of wildlife management in more detail and gives further descriptions.

49. This is discussed in detail by S. J. Riley et al., 2002, "The essence of wildlife management," *Wildlife Society Bulletin,* 30(2), 585–593; H. Harju, 2003, "Perspective: Politics is hurting wildlife management," *Casper Star Tribune,* April 14, A10; B. Manning, 2003, "Director's opinion," *Wyoming Wildlife News,* May–June, 2. See also T. W. Clark, M. B. Rutherford, M. Stevenson, and K. Ziegelmayer, 2001, "Conclusion: Knowledge and skills for professional practice," 253–276 in T. W. Clark, M. Stevenson, K. Ziegelmayer, and M. B. Rutherford, eds., *Species and ecosystem conservation: An interdisciplinary approach,* Yale School of Forestry and Environmental Studies Bulletin No. 105; Clark, *The policy process: A practical guide for natural resource professionals.*

50. This process was reviewed by G. Wilkes, 2002, "Sage grouse plan is a compromise," *Pinedale Roundup,* August 8, 17. The process built trust and cooperation between varied interests, according to Linda Baker, activist, and Al Sommers, rancher, pers. comm., March 2002.

51. Improving natural resource decision making is discussed by R. D. Brunner et al., 2002, *Finding common ground: Governance and natural resources in the American West,* Yale University Press, New Haven.

52. These recommended standards embody democratic decision making, according to H. D. Lasswell (1971, see endnote 2). Furthermore, Lasswell recommended that decision making should (1) strive to arrange for common interests to prevail over special interests, (2) give precedence to high-priority rather than low-priority common interests, (3) protect both inclusive and exclusive common interests, (4) in the face of conflicting exclusive interests, give preference to the participants whose value position is most substantially involved, and (5) allocate values of sufficient magnitude (power, money, knowledge, skill, and so on) to enable authority to be controlling.

53. Common pitfalls in decision making are well documented. A brief listing of these is given in Clark, *Averting extinction: Reconstructing endangered species recovery,* 171. These known pitfalls warn us to avoid simplistic problem definitions with a single objective; not to let a single organization dominate; to balance local and central (usually government) participation; to bring in diverse experts right away and minimize their biases; not to let expert opinions overrule the preferences of those affected by decisions; not to focus solely on easily measured and easily understood data but to ask the hard questions; to coordinate government decision making; not to allow agency rivalries to block achievement of the goals; not to overcontrol participants; to avoid intelligence failures and delays; to make sure that government management agencies do not succumb to the natural selfishness of the private sector; to ensure adequate appraisal; to ensure that decision makers are sensitive to criticism; to learn explicitly from experience; to deal fairly with people; and more.

Part Two: Cases

Chapter **3**

Mountain Lion Management: Resolving Public Conflict

Tim W. Clark and Lyn Munno

In February 1999 a female mountain lion and three kittens took up residence on Miller Butte on the National Elk Refuge barely a mile from the Jackson, Wyoming, town limits (figure 3.1). Over the next 42 days nearly 15,000 people came to see them. This mountain lion family became the focus of great public interest, and its brief tenure in this highly visible location tapped a huge latent interest in wildlife watching and, subsequently, in the welfare of mountain lions. The media covered the story for months and continues to report on mountain lion management, a subject given little attention in the past. Many photographs were taken and a book was written on the refuge's mountain lions.

Following closely on this unprecedented public expression of wonderment and delight in the lion family, in June 1999 Wyoming Game and Fish Department (WGFD) announced that it intended to increase the mountain lion harvest for the hunting area encompassing Jackson from five lions the previous year to twelve. Within the wider area covered by this book (about 22,000 square miles in western Wyoming, south of Yellowstone National Park—see figure 1.1), the quota increased from 20 to 36 mountain lions along with an unlimited quota for males in Star Valley (36 miles south of Jackson). The department gave little explanation or justification for increasing the hunting numbers so significantly in one step. Newspaper editorial and article headlines over the next couple of years tell the rest of the story: "Public weighs state plan to increase cougar hunting," "Lions should be

Figure 3.1a
Female mountain lion and her three large cubs on Miller Butte, National Elk Refuge, near Jackson, Wyoming, March 1999 (photo by Tom Mangelsen).

subject to better management," "36-lion hunting season sought," "Lion-hunting opponents attack Game & Fish," "Lion hunt justified," "Game and Fish hears lion policy criticism," and "Game and Fish ignores mountain lion science."[1] The magnitude and timing of this management policy change, coupled with the department's failure to involve the public meaningfully, created a rift between WGFD and many citizens, particularly in Teton County. This incident set the contentious tone for mountain lion management that persists today.

In this chapter we examine the causes and consequences of conflict over how to manage mountain lions in Wyoming, and we offer ways to find common ground. This chapter (1) briefly describes the natural history, populations, and management of mountain lions, (2) analyzes how the management process became politicized, and (3) offers options to improve management. We have followed the mountain lion management issue since 1999. We attended WGFD public meetings in Jackson, talked with people on all sides of the issue at these meetings, continued discussion outside meetings, followed the issue in newspapers, surveyed the scientific and management literature, and talked with mountain lion researchers in Wyoming and Idaho. In

Figure 3.1b
Citizens and photographers viewing the Miller Butte mountain lions on National Elk Refuge near Jackson, Wyoming, March 1999. Nearly 15,000 people came to see the lions in a 42-day period (photo by Tom Mangelsen).

March 2002 we interviewed approximately 40 people from WGFD, the U.S. Forest Service, the ranching community, hunting outfitters, and conservation groups.

Natural History, Population Dynamics, and Management History

Mountain lions have not received as much attention from the public as wolves and grizzly bears, in large part because of their natural history. To appreciate the species' conservation needs and to understand the conflict as well as the options for improving management, it is important first to look briefly at lion ecology.

Natural History[2]

Mountain lions (*Puma concolor*) are the second largest felid in North America, slightly smaller than the jaguar (*Panthera onca*). Mountain lions are muscular and slender with a long tail that is one-third of the animal's total length. Males are larger than females, though size differences are often difficult to distinguish except at close range. Although males tend to weigh more than females, there is usually little difference in length. Males weigh between 115 and 150 pounds and stand 22–31 inches high at the shoulders. Females weigh 75–110 pounds and stand 21–30 inches high at the shoulders.

Mountain lions have a large range across western North America and extend from Panama to Canada. There is also an endangered subspecies, the Florida panther, present only in Florida. After the near elimination of mountain lions from the United States in the first half of the 20th century, protective laws beginning in the mid-1960s allowed them to reestablish or increase their populations across the West.

Mountain lions typically occur in topographically varied habitats. Their rugged habitat is one of the main reasons that lions are seldom seen. They live in areas with high prey densities and enough vegetation and topography for good hunting cover. Maternal females usually select dens in rock outcrops, dense shrubs, or under conifers well out of sight of people.

Mountain lions spend most of their time alone. Scientists classify them as noncooperative rather than solitary hunters. Their social organization is a territorial "land tenure system" wherein resident adults hold exclusive dominance. Adult males actively defend their territories, and thus the territories are mutually exclusive. Females do not defend their home ranges, so their use areas typically overlap. Independent subadults still too young to breed usually emigrate out of their natal areas and establish residency elsewhere. Subadults usually disperse when 10–22 months old. Male offspring disperse from their maternal home range, whereas the behavior of young females is more variable; sometimes they emigrate and sometimes they set up ranges on the edge of their mother's range. This dispersal pattern of young lions has important adaptive purposes, including maintaining genetic diversity and replacing populations that have been disrupted or diminished.

Mountain lions are both polygamous and promiscuous, that is, adult males breed with a number of different females. Females reach sexual maturity at 24 months and breed with more than one male during an estrous cycle. Gestation ranges from 82 to 103 days and litters average 2.6 cubs (range 1–6) with an interbirth interval of 17–24 months if the mother has successfully raised her cubs to at least 1 year. Cubs (or kittens) are born at any time of year, but there are different birth pulses ranging from June to September in the southwestern states and from August to November in Wyoming. Female lions are solely responsible for rearing cubs. Cubs rely on mother's milk for the first 2 months and then begin to accompany their mother on hunts. They remain with their mother until they are about 1 year old.

Population Dynamics

Mountain lions occur throughout western Wyoming, but little is known about their populations in scientific terms. Because of their elusive nature, they have been very difficult to study in the field. The only data in northwestern Wyoming come from long-term studies done in northern Yellowstone National Park by scientists at the Hornocker Institute. In addition, John Laundré and his colleagues have been working on another long-term study in a nonprotected area in south central Idaho, and Kenneth Logan worked on population dynamics in both the Big Horn Mountains and in New Mexico. Mountain lion research is underway in Teton County, originally undertaken by the Wildlife Conservation Society and more recently by Beringia South researchers.

The scientific literature gives us our current picture of lion population dynamics. Mountain lions typically show "density dependence" and occur at fairly low densities. The reason for this is their territorial behavior, particularly in males, and required prey density. It is unclear which of these factors has a stronger influence on determining regional carrying capacity.

There are three age classes—cubs or juveniles (birth to 12 months), subadults, and adults. Once mountain lions reach 1 year old, they begin to disperse; this dispersal period between 12 and 24 months is considered the subadult class. Mountain lions live on average until 10 years of age. Survival rates seem to be constant during these years. Taking survivorship into account, a female has an average of three to four female offspring in her lifetime.

Based on mark/recapture and radio telemetry studies, mountain lions exist in densities of 0.6–2.2 resident adults or 1.4–4.7 total mountain lions per 100 km^2 (about 63 square miles) in nonhunted populations. In Idaho the average minimum density of mountain lions was

0.77 resident lions per 100 km^2 over an 11-year period, and the densities fluctuated widely over that time.[3]

In almost all populations, human-caused mortality is the leading cause of death. Legal hunting accounts for the majority of deaths in most populations. However, even in protected populations, predator control, illegal hunting, and vehicle collisions cause many lion deaths. Survival rates for mountain lions vary depending on habitat quality and hunting pressure.

Mountain lion populations fluctuate based on prey availability as well. According to both WGFD and John Beecham (formerly of Idaho Fish and Game Department and later the Wildlife Conservation Society), healthy mountain lion populations in the 1990s were largely the result of healthy elk and mule deer populations during that time. Elk numbers currently are quite high in western Wyoming and provide abundant prey. Elk are fed on state-operated feedlots during winter and are concentrated in much higher densities than under natural conditions, making herds very susceptible to disease, including chronic wasting disease and brucellosis. These herds already have brucellosis, and there is currently considerable public concern that an outbreak of chronic wasting disease could drastically decrease herd sizes. Mule deer populations were high in the early 1990s, but have been decreasing for a variety of causes. If the elk population does decline and mule deer numbers continue to go down, the lions' food base will be much reduced, which will affect survival rates for all ages and sexes. Mountain lions also consume other ungulates occasionally, including moose and livestock. They are primarily carnivores, although at times they scavenge animals that have died from other causes. On rare occasions they eat grass to get roughage to remove parasites or to expel ingested hair.

Management History

Mountain lion management is solely the responsibility of the State of Wyoming, specifically WGFD. They have been managed by the state since statehood in 1890, even though many mountain lions live on federal lands. Current management policy is a product of history, the perceived status of lions, the values of managers and citizens, and state policy preferences. In Wyoming today mountain lions are classified as trophy game animals, and hunting quotas establish the number that can be killed in a given year and location.

Prior to colonization, mountain lions had the widest distribution of any carnivore in the Western Hemisphere, throughout both North and South America.[4] As nonnative people began to settle the West in the 1860s, mountain lions, along with other carnivores, were considered a threat to safety and livestock, and people sought to eliminate them from the landscape. A bounty was placed on them. As a result, thousands of mountain lions were killed across the West, including in Wyoming. In the eastern United States, they were eliminated from their entire range except for a small population in southern Florida.

In the 1970s Americans' attitudes toward wildlife and carnivores greatly changed. More people began to view carnivores as an asset and part of the country's cultural and natural heritage. In 1973 the Endangered Species Act was enacted, and that same year Wyoming instituted the first mountain lion hunting season for which a license was required. In the 1970s hunting harvest limits for mountain lions were set in all states except Texas (where an unlimited hunt still exists). California is currently the only state in the West that has banned the hunting of mountain lions.

As a result of these policy changes mountain lion numbers have been increasing in most areas across the West. There are few data available on the actual number of mountain lions in western Wyoming. All estimates are based on anecdotal sightings and hunting success rates. Although there is certainly disagreement as to the magnitude of population increases in the area, it is likely that lion numbers have increased since historic lows in the 1970s. As mentioned earlier, mountain lions are density-dependent and will eventually stabilize at a carrying capacity for the region. However, because of past human influences, mountain lions were most likely kept far below that capacity in western Wyoming. The regulated hunting since the 1970s has likely allowed the populations to begin rebounding. However, current hunting quotas are suspected by some in the public to be too large. Recent data from the entire western United States suggest that lion populations fluctuate on a cyclical basis in response to deer numbers, which are largely regulated by weather conditions.[5]

This natural rebounding, coupled with high prey densities in the 1990s and more people using the backcountry, is most likely the reason that people have sighted more mountain lions in the last two decades. In 1999 WGFD decided that the increase in sightings justified higher hunting quotas across the state. They doubled the quotas in most areas and raised the number of mountain lions that could be hunted in Teton County (Hunt Area 2) from 5 to 12.[6] This hunt area was subdivided in 2003 and the quota was increased further, with the harvest apportioned equally over both areas.[7] The total mortality quota for Area 2 and the newly created Area 29 increased by four in 2003 compared to the original Area 2 in 2002, although the female quota remained the same. During this time, public hearings were held to hear testimony from proponents and opponents of the quotas. The Wyoming Game and Fish Commission voted nevertheless to maintain the quotas. Mortality quotas established for 2004 will remain in effect through 2007 unless conditions warrant a change in the view of the department.

Since there are few data on mountain lion population densities and numbers or the impact of increased hunting, it is unclear how these policy changes have affected the populations. Representatives from WGFD feel that the populations are healthy and can easily tolerate increased hunting mortality. In its 2001 annual report the agency states that "annual harvest estimates have also been increasing, which appears to indicate healthy and expanding mountain lion populations in Wyoming and other western states." This view was also expressed during our interviews with state officials in 2002 and 2003 and at public meetings. Mountain lion harvest data are reported for the entire state and by region.[8] For the state, total take was 173 in 1998, 208 in 1999, 186 in 2000, 214 in 2001, and 201 in 2002. In the western part of the state, 56 lions were taken in 1998, 85 in 1999, 84 in 2000, 85 in 2001, and 89 in 2002. The number of hunter days for this same region was 181 in 1998, 388 in 1999, 269 in 2000, 262 in 2001, and 358 in 2002. Within our area of focus, which includes Hunt Area 2, hunting quotas have been met each year. The same is true for hunt areas to the south. The number of illegally killed or poached mountain lions is unknown.

The exact impact of hunting and how management can and should respond to fluctuating population numbers is still in dispute. In a June 2001 article in the *Jackson Hole Guide,* Bernie Holz, wildlife supervisor of WGFD for the Jackson-Pinedale district, stated that "the past two hunting seasons have stabilized or reduced lion populations in

Jackson and Star Valley. The agency has seen a shift in hunter kills from large adult cats to smaller younger lions which is a sign of an 'exploited' population."[9] For the 2001/2002 hunting season, WGFD placed female subquotas in many regions and eliminated the unlimited hunt in Star Valley.[10] For the 2002/2003 hunting season, once six females were killed in Hunt Area 2, the hunting season was over for the year. However, this quota was actually higher than the number of females hunted in the previous several years, giving many the impression that this was a disingenuous change. Some researchers and local citizens feel that mountain lions are being overhunted throughout the American West.[11]

Managing mountain lions is a matter of human safety as well. News articles attest to people's fear of lions and their concerns about safety. For example, in Hoback Junction at the southern end of Jackson Hole, a woman shot and killed a mountain lion that had killed her cat.[12] She blamed her neighbors' feeding of mule deer for drawing the lion into the area. A few articles in the *Casper Star Tribune* tried to address concerns about safety after mountain lions were sighted in that city in 1999.[13] *Cat Attacks* is a compilation of stories of people who have been killed by mountain lions.[14] The truth is that no mountain lion has ever killed a person in the state of Wyoming, and in general, the risks are very small. In Montana a survey was conducted to assess people's tolerance of mountain lions and the perceived risk to personal safety.[15] It was found that people's perceptions of risk were far higher than the actual risk. However, it is still an important management issue that should not be left out of the policy discussion.

Politicization of Management

The mountain lion management process is thoroughly politicized. Citizens have expressed a wide range of concerns at public meetings and in newspapers. Some people support current management approaches, whereas others want decision making and quota setting to be changed or made more rational by basing them on scientific data. People who oppose WGFD policies point to the lack of data supporting WGFD management policies, the hunting of females which may leave dependent young abandoned to die, and the lack of meaningful public involvement in the management decision process.[16] At least three factors have led to the ongoing controversy—the values at stake, the goals of management, and the role of the public in the management process.

Values at Stake

Values are at the heart of the conflict over mountain lion management. A good example of this is described in a letter written by Lloyd Dorsey, then the Jackson field representative of the Wyoming Wildlife Federation, to the director of WGFD in 2001 complaining about the political maneuvering in lion management. Dorsey's letter supported Joe Bohne, a WGFD wildlife management coordinator whom the department had abruptly transferred from the Jackson/Pinedale region to another part of the state. Dorsey wrote: "[He] has fallen victim to the effective lobbying from livestock producers, big game outfitters, and Department politics and been transferred to Lander. This was a completely unexpected and unwilling transfer. One of the allegations against [him] was that he was too friendly with the enemy, meaning the conservation community [including his] stance and willingness to discuss responsible management alternatives [and to] tell the truth about predator control and mountain lion hunting."[17]

Another example of values and politics is evident in WGFD's justification of the financial dimension of mountain lion management. A common argument heard from hunting outfitters and WGFD employees is that the hunting community should have more say in mountain lion management because it bears the cost of the state's wildlife management through the purchase of hunting licenses. This rationale favors hunters over all other people. That buying hunting licenses should privilege hunters in decision making within public agencies is disputed by many people. In addition, according to the WGFD 2001 report, mountain lion hunting licenses brought in $68,450 to the agency in 2000 (although this does not include additional benefits that accrue to the state from lion hunting, such as tourist revenues from hunters), but maintaining the lion management program cost $540,901.[18] Hunting mountain lions is clearly diseconomic, if these data are representative. Hunting is an important part of the state's cultural heritage, and in this case a commitment to continued hunting seems to override economic and other legitimate considerations.

These value differences and how they come together in mountain lion management have led to conflict. The clash of values is clear in newspaper headlines, such as: "Lion hunters display the courage of Pooh," "Lion hunt justified," and "Game and Fish ignores mountain lion science."[19]

How did lion management become so politicized and why? It is reasonable to assume that people will behave in ways that they perceive will leave them better off than if they had behaved in other ways. Sometimes, in fact, people's actions seek to satisfy personal needs that are unconscious even to themselves. It is the competition over whose values will dominate that has politicized the process. In this competition, values such as power, respect, and rectitude are requisitioned by people on all sides to advance their demands. WGFD has made decisions that ally it with localist values over those of the New West, creating a decision process that appears to be driven by special interests and excluding legitimate interests that disagree with the state's management policy. Whereas WGFD benefits from the current management regime in terms of power and rectitude, the agency has lost respect from many in the community, not to mention monetary losses. The way the department has gone about its decision making has prevented it from achieving a broadly supported, common interest outcome.

The value dynamics play themselves out through the institutional system of wildlife management, which has a certain structure and rules of engagement (see chapter 7). The present institutional system of wildlife management did not just spring up spontaneously. It reflects the views of the most powerful interests, in this case WGFD, which largely controls the institution and sets many of the boundaries and content of the public discourse about mountain lion management. It determines who will be heard and taken seriously at public meetings, for example. This history of the issue shows that present institutional arrangements are ill equipped to take both groups—Old West localists and New West environmentalists—into account simultaneously in setting lion management goals and implementing programs. The institutional system, its structure, operation, and leadership, is unable to clarify, secure, and sustain a common interest in lion management policy. That the mountain lion issue is mired in conflict is one measure of how this system functions to manage the politics and find common ground. The politicized situation leaves many citizens feeling that they have no meaningful role or ability to effect change in the management process. Alienation of significant segments of the public only politicizes management further and draws down any trust or good will.

Goals of Management

Because mountain lions were never extirpated from the area and because of their secretiveness, these carnivores have kept out of the limelight and out of high stakes politics until recently. This situation contrasts dramatically with grizzly bear and wolf management, which has received intense public attention since their return to parts of the southern Greater Yellowstone Ecosystem. This pattern changed for mountain lions in 1999 when the lion family appeared on Miller Butte, creating enthusiastic fans and, ultimately, conservation-minded supporters, both locally and nationally. As Ines Rukovskis of Jackson stated, "Every day people would wait for hours in the cold to catch a glimpse of the mother and her cubs in all their aliveness and beauty."[20] Years later, people from all sides of the issue still talk about the Miller Butte mountain lions as an amazing, once-in-a-lifetime experience.

Goals establish targets for action. They reflect what we value. Often they represent the dominant interests at stake. In Wyoming the authoritative, strategic goals of mountain lion management are unclear, as are the instrumental goals. Nevertheless, management carries on and harvests continue. In 1992 a committee was set up to write a management plan. Fourteen coauthors from WGFD and the University of Wyoming produced a draft plan in 1996, which began by noting that public interest in lion management has steadily increased in the last 20 years. The draft plan identified five public viewpoints on mountain lions: (1) as rare and in need of protection, (2) as a prized hunting trophy, (3) as useful for training hounds, (4) as a magnificent wild predator, and (5) as a threat to livestock and big game herds. The draft plan did not, however, commit the state to the overriding goal of finding common ground among these five interests or meeting the standards of "best" lion management. In addition the draft plan focused more on designing a planning means for the future than on setting goals and priorities for the present.[21]

Although this draft plan was not comprehensive in terms of an overall management goal for mountain lions, it did establish specific, concrete ways that the agency planned to approach mountain lion management in the future. Its first recommendations were to establish an effective public input process to involve a "broader base of constituents at the local level regarding mountain lion management" and to "conduct periodic statewide and regional surveys of Wyoming residents to determine broad-scale and local public attitudes to assist future mountain lion management decisions."[22] These recommendations were not effectively implemented. The draft plan called for improvements in tracking the status of the species as well as more systematic population projections to gain a better understanding of population dynamics. Based on WGFD's failure to provide evidence of its tracking activities or population models, we can assume that effective action was not taken on this recommendation either.

The draft plan was never adopted because soon after it was written WGFD decided that mountain lion numbers were increasing and therefore the agency did not need a plan.[23] Now, some years later, WGFD is managing mountain lions without a formal, written agency plan. This lack of clear goals and a publicly acceptable plan or a means to get them leads to distrust about the agency's decisions on lion hunting quotas, population viability, and other issues. It also means that mountain lions are being managed on a short-term basis without any overall, long-term strategy that can be justified to the public.

Although mountain lions are resilient animals, their population numbers may fluctuate over years based on variations in management policy and ecological conditions, and it is unclear how officials take this variability into consideration. If they are overhunted, if disease in the elk herds increases, or if mule deer numbers continue to decline, the public's concern for the mountain lion's future will be heightened. Although WGFD has stated that it would reduce hunting quotas if problems arise, there is currently no policy to establish a target population size for the southern Yellowstone region. Because WGFD management is vague and there are few data on population dynamics in the region, it is unclear how the agency's management policy might change in the future as conditions change.

Furthermore, mountain lions prey on game animals that are important to the state. For example, elk hunting is an important food source for people and an important source of funds for WGFD. Elk management—from hunting quotas to maintaining elk on feedlots—is widely discussed and debated. Currently, elk numbers in Wyoming are quite high, and WGFD has been trying to lower the numbers over the past few years. Some people, outfitters and hunters in particular, feel that predators are reducing elk populations. Predator-prey relationships are very complicated, and it is therefore understandable that it is difficult for WGFD, through its management policies, to "balance" predator numbers with prey numbers to the extent that may be desirable. However, it seems that the department uses this predator-prey dynamic as a justification for measures designed to appease the dominant elk hunting public, but not other interests, including lion conservation. According to WGFD officials, the increase of mountain lion hunting quotas was not intended to increase numbers of elk. As stated in its public announcement in the *Jackson Hole Guide,* "The Wyoming Game and Fish does not believe that our prey species are being suppressed by the current mountain lion populations and consequently, does not support raising mortality quotas to bolster big game populations."[24] At the same time, a WGFD official stated during an interview that the agency does raise mountain lion hunting quotas if outfitters in a region indicate that they are concerned about mountain lion impacts on elk. He stated that it usually only takes a marginal increase in harvest numbers to satisfy the hunting outfitters.[25] The contextual basis for goals and plans is complex, extending beyond the biology of the animals and resting firmly on the human values at stake and other factors.

The Role of the Public

In the case of grizzly bears and wolves, there are multiple agencies involved with many different agendas, which sometimes coincide and sometimes compete with one another. Since mountain lions are managed by WGFD alone and because there is no federal mandate for their protection, the number of participants involved in the process is much lower than in management of grizzly bears and wolves. It seems that WGFD's approach to lion management is to be unwavering in its policy making and always to favor hunting interests. This gives the impression to many that public input, especially from those who question WGFD's management, will not influence state policy. As WGFD stated in its 2001 newspaper announcement, policy is not based on "public opinion."[26] WGFD's actions provide evidence of the agency's strong support for states' rights and for its sole jurisdiction over policy making. The institutional system for managing carnivores also embodies the agency's interests.

The public became concerned over the state's lion quota increase in June 1999. The most vocal opponents of the policy were Tom Mangelsen, a world-renowned wildlife photographer, and Cara Blessley Lowe, a filmmaker and author. They were concerned about the lack of data on mountain lion numbers, the hunting of females when dependent young may be abandoned, the lack of knowledge about potential long-term effects of WGFD policy, and the lack of public involvement in decision making.[27] Mangelsen became interested in mountain lion management, in part, after filming the lion family on Miller Butte. He went on to coauthor a book with Lowe about these lions called *Spirit of the Rockies: The Mountain Lions of Jackson Hole*.[28] Others in the community have also spoken out against WGFD management policy in numerous letters published in Jackson Hole newspapers.[29]

As a result of this discord, WGFD convened a public meeting in April 2000 in Jackson.[30] It was an "open house" where people could express their opinions about the current policy; all were given an opportunity to speak. But many people were frustrated because they believed that the state's management decision was a forgone conclusion. Many also felt that the meeting was merely "window dressing" to appease vocal opponents in the public, that WGFD conducted the meeting merely as a formality for people to vent their opinions. Finally, as indicated by newspaper letters and opinion pieces, in general people felt that they did not have a chance to be heard, to be taken seriously, or to make a difference. This public sentiment only added to the disharmony.

Later, in the spring of 2000, Mangelsen attended a meeting of the Wyoming Game and Fish Commission to offer verbal and written testimony. Identifying himself as a hunter and long-term Wyoming resident, he called into question the quality of the science the state was using to set quotas. He also questioned the anecdotal accounts of conflict with humans unsubstantiated by scientific data, the relations to prey species, the accidental killing of females with cubs, the ethics of methods used to "tree" mountain lions, and other aspects of the state's plan. He was later told by a state official that his comments were disregarded because he was not representing the hunting interest.[31]

In a newspaper opinion piece, "Mountain lion hunters show the courage of Pooh," Jack Turner, an elk hunter and long-term Jackson Hole resident, stated that mountain lion hunting is "condescending, self-serving, shameful, cowardly, unmanly, unsporting and chicken. Why do we hunt lions? That's a no-brainer—money."[32]

At a second WGFD public meeting in June 2001, the department only allowed people to write comments on flip charts scattered around the room. There was no opportunity for a public airing or public discussion of management in general or the hunting issue in particular. This led to more public criticism. As Tom Mangelsen wrote after this meeting, "I again learned the extent to which Game and Fish management is insular, not based in science and genuinely unconcerned about public opinion. Even though the public was invited to write down comments, the agency is not required to consider all these comments. Rather, it takes into consideration only those ideas or opinions that are aligned with its current approach to management."[33]

WGFD officials also viewed these public meetings as unsatisfactory because they felt that the public showed up simply to berate the agency's actions. The agency seemed to view the meetings as a opportunity to justify its current policy to the public, not as a way to engage the public to help shape current and future policies. As one official said, WGFD wanted to "lay out on the table for folks the how and why of mountain lion management" and "simply show people how we approach lion management in Wyoming and here locally."[34] This meeting's design was a combination of what is called "passive participation" and "participation in information giving,"[35] wherein people are told what has happened and what is going to happen. Communication was predominantly unidirectional from state officials to the public audience. The public resented this arrangement.

Another critic of WGFD policy and its public meetings was Rebecca Rundquist, a lawyer living in Jackson. "[WGFD's] most recent June 18th public meeting . . . evidenced a regression in any move towards public dialogue. One can not have a public conversation with a magic marker and a large pad of paper on an easel. Physically, the tenor of this meeting was set up to be divisive. I quickly realized this was not going to fit any typical, democratic definition of a public meeting. . . . The agency's explanation of 10 identified issues surrounding mountain lion management focused primarily on self fulfillment of its preexisting standpoint instead of thoroughly examining alternatives that may better address species recovery in its current context."[36] Other people in attendance voiced the same concerns both to WGFD and to each other. Although Mangelsen, Lowe, and Rundquist were the most vocal, they were not the only members of the public who viewed the WGFD meetings and management policy with suspicion.

Both meetings were destined to fail because the expectation of the public was to influence and help guide mountain lion management, while the expectation of the state was to stick strongly to its own course and preserve its autonomy while at the same time making the public better understand and support its decisions. Many people learned from this experience. For some it was their first involvement in a wildlife management issue. They learned that the process is highly political and that public input, especially from people with values and interests different from those institutionalized in the state, does not influence WGFD management policy. It is not only the public that has been disappointed in these exchanges. The agency also recognized that mountain lion management was a public relations nightmare. As a result, WGFD has been disinclined to hold more public meetings on lions. The state felt that it had spent considerable effort trying to explain its position on the issue, for example, with its full-page ad explaining its rationale in June 2001 in the local Jackson paper.[37] Yet the agency's efforts to assuage its opponents were unsuccessful. WGFD treated the controversy as though the problem were the public's lack of information. The agency's solution was to explain its policy to the public, assuming that once people had this information they would support the state's policy. But the problem was, in fact, not lack of information, but strongly differing human values. People wanted their concerns to be taken seriously in deliberations on mountain lion management, not summarily dismissed. After years of trying to be heard, Mangelsen felt there was no way to effect change in the state's policy. He and Lowe formed a nonprofit organization, The Cougar Fund, to promote mountain lion conservation on a national scale and to work with people on lion conservation in the West.

In 2003 a similar drama played out over the subdivision of Hunt Area 2 into two units and the increase of the quota from 12 to 16 mountain lions annually. This policy change was enacted by the Wyoming Game and Fish Commission in Sheridan at their late July meeting. The commission approved the new season and a change in the management review period from once a year to once every 3 years "with almost no discussion and [it] did not publicly address cougar supporters' concerns."[38] Franz Camenzind, executive director of the Jackson Hole Conservation Alliance, one of several Teton County citizens who drove nearly 6 hours one way to address the commission, said, "This whole thing is being driven by hunters and outfitters that cash in on killing cougars."[39] Tom Mangelsen said, "[The Commissioners'] minds are made up long before anybody enters the door."[40] John Emmerich, a WGFD staff member, agreed with Mangelsen, adding, "We don't know how many lions we have," and noting that agency staff are, nevertheless, confident that lions are thriving in the state and can sustain increased hunting.[41] After returning from the meeting, Jackson resident and lion photographer Tim Mayo wrote in a letter to the editor, "The 'public hearing' was insincere, horribly disorganized, exhibited a constantly changing agenda and an outcome arrogantly and unquestionably predetermined."[42]

The ranching community has been very vocal in its opinions about the management of wolves and bears, but it has not been nearly as involved in the debate on lion hunting quotas or lion management overall in the state. After interviewing members of eight ranching families in the region, we discovered that these ranchers do not hold as strong views about mountain lions as they do about other carnivores. Although some ranches have experienced livestock losses from mountain lions, they are much less concerned by these depredations, in part because mountain lions have always been a part of the landscape, and in part because of the elusive and infrequent manner in which lions prey on livestock. In addition, because mountain lions are a hunted species, ranchers feel that they have control over lion problems. Also, if they suffer losses, they can claim damages from the state.

Hunting outfitters receive around $3,000–3,500 per hunted mountain lion, so the financial benefits to them of doubling hunting quotas in the state were quite high.[43] Also, some hunters favor the lion hunt because they feel that it reduces predation on elk and deer. However, many oppose the use of radio-collared dogs to track mountain lions and even the hunting of carnivores in general. Interest in hunting mountain lions has increased since the quotas were raised, and at this point if quotas are decreased significantly it is likely to cause more conflict. Smaller, incremental changes are more likely to be accepted in the community rather than large year-to-year shifts.

Although all these people have different values and viewpoints about how mountain lions should be managed, the differences are not necessarily irreconcilable. Although conflict currently exists between WGFD and some citizens, in Teton County especially, in general the public across the region are relatively neutral about mountain lion management. One interesting point is that both hunters and nonhunters and both proponents and opponents of the current policy consistently express a lack of understanding of how and why WGFD sets its hunting quotas and its management policies in general. The situation could be significantly improved if contending parties would discuss the basis for adequate management policy, and agree on that basis, appropriate goals, and the kind of adaptive management needed. What was and is still needed is a WGFD management decision-making process that is genuinely inclusive, one that invites participation in discussion, problem solving, analysis, and production of a publicly acceptable action plan. A new approach, an upgraded version of the public participation called for in the draft mountain lion plan, is needed.

Management Options

We have two goals in recommending alternatives to improve mountain lion management. The first is to alleviate conflict and create a decision-making process for management that is clear, fair, justifiable, and based on responsible public and agency input. The second goal is to manage for mountain lion viability over the long term and to establish consistent management policies that serve common interests. We believe that it is possible to develop a policy that is acceptable to most parties. The problem at present is the lack of an effective decision-making process that might clarify and secure the common interest. WGFD's approach has alienated many people and left others feeling uncertain about the integrity of the state's policy decisions.

If the status quo continues, we project that tensions and open conflict will persist and mountain lion populations will perhaps suffer. We recommend four alternatives that should be addressed simultaneously: (1) establish a formal mountain lion management plan, (2) produce that plan through an effective, public, problem-solving process, (3) work together with all interests to protect habitat and prohibit wildlife feeding, and (4) use a metapopulation approach to mountain lion management and hunting. All of these will require that WGFD improve its interactions with the public and use state-of-the-art management tools.

Adopt an Official Management Plan

We recommend that WGFD establish a formal management plan for mountain lions, clearly establishing goals for population sizes, habitats, and harvests, as well as public input and education, for the short and long terms. The plan needs to substantiate how the department will establish quotas, including how it will collect data on population abundance and distribution, and if information is lacking, on what basis it will decide on acceptable quotas. The plan should also establish how the agency will manage mountain lion populations, in cases when numbers are increasing as well as when they appear to be decreasing. A management plan will not only clarify goals for the public, but it will also be very helpful for the agency itself. A formal plan supported by a broad public would allow WGFD to give a clearer message about its intentions and allow it to make better on-the-ground management decisions over the long term.

Genuinely Involve the Public

For mountain lion management to be successful, it needs broad-based public understanding and support. We recommend that WGFD initiate small, well-crafted, cooperative, problem-solving meetings with a wide variety of people to build common ground. The process should demonstrate high standards in order to develop a reputation for honesty and fairness. Personal diplomacy will be needed to bridge the hostility, resentment, and ill will spawned during past conflicts. This new process should exhibit what is called "interactive participation," that is, it should include genuinely cooperative, joint goal setting, analysis, fact finding, interpretation of data, and recommendations. It should be problem-oriented and contextual. Finally, it should involve interdisciplinary methods that seek multiple perspectives, draw on both natural and social sciences in an integrated way, and make use of systematic and structured learning processes.[44] This in turn can lead to action plans and the formation of working relations that are scientifically rational, politically practical, and morally justifiable.

Community-based problem-solving approaches are popular, but they must be well organized and they must seek common interest outcomes.[45] Finding responsible people to participate in community-based problem solving should be an open and fair process. Self-selected participation should also be permitted. Participants should be committed to improving the rationality of the process rather than vying to control the outcome to achieve their special interests. Their interests should be both valid and appropriate. Many authors have described community-based problem solving, its pitfalls and benefits, and offered practical designs (see chapter 6).

Overcoming any WGFD resistance is essential to this recommendation. After public meetings in 2000 and 2001 that the agency itself considered unsuccessful, the state concluded that it would not hold future meetings. This decision makes it difficult to implement our recommendation. Some people in WGFD are aware that the mountain lion case was not handled well, but there appears to be no formal effort to seek creative, workable alternatives that might move the process beyond the current impasse. Since lack of an adequate public process has led to the highly politicized situation that exists currently, getting the process turned around and operating effectively would to do much to defuse the politicization. We urge the state to reconsider holding public meetings and to strive for genuinely participatory forums that engage people's interest and good will and build trust.

A new lion management process that genuinely involves the public in interactive participation should strive to meet widely recognized standards of public policy making, such as timeliness, inclusivity, and respect. *Timeliness,* for example, is critical for obvious reasons. Planning, decision making, or information gathering (including scientific research) should not drag on for years without taking action to solve problems. Delays lead to resentment, anxiety, fear, and exacerbation of existing problems. *Inclusivity* is a fundamental principle of democracy. The people who are affected in any way by an agency's decisions or anyone who has something to contribute should have a say in the process. *Respect* means being taken seriously as though one's views mattered. And mutual *respect*—a pattern of deference that takes the form of listening, participating in dialogue, and fully and genuinely engaging one another in problem solving—between government agency workers and the public will go a long way in developing a successful process. Other standards to be achieved include reliability (factual), fairness, and amelioration (reducing hostility).

Protect Habitat and Prohibit Feeding

Currently, there is such animosity between WGFD and concerned citizens of Teton County that it is difficult for either side to see that there are opportunities for cooperation. Although citizens should eventually work with WGFD on hunting quota issues, starting with issues that are less contentious may be a better way to create a successful pattern of interaction and to build trust. Both WGFD and people outside the agency share an interest in reducing the number of "management removals," that is, mountain lions that are killed because they have gotten into trouble with people. As development in Teton County and elsewhere has moved into prime wildlife habitat and as more people feed deer and thus draw lions into populated areas, there have been more conflicts. A citizens' group as well as the Teton County Commission worked on this issue, and in 2003 the commission passed a feeding ban. This is an issue on which the citizens who are concerned about mountain lion protection could work side by side with WGFD and other groups on a common goal and begin to develop trust.

Manage Mountain Lions as Metapopulations

Management of mountain lions, including lion hunting, should be approached on a "metapopulation" basis. John Laundré and Tim Clark, based on 16 years of research in Idaho, have outlined how such an approach could work across the western United States.[46] Cara Blessley Lowe, one of the citizen activists, has also recommended a metapopulation approach to mountain lion hunting in western Wyoming.[47]

The metapopulation approach to lion management, based on the "source/sink" concept in population ecology, would designate some areas as "sources" closed to hunting and some areas as "sinks" open to hunting.[48] Mountain lions can disperse over large areas and, therefore, except where isolated by habitat constraints, the lions in distinct areas often act as one population. Protecting some areas as sources, then, would help to protect populations of mountain lions in the long term, despite the difficulties of counting the number of lions present in an area or the threats of random events such as drought or prey decline. These source areas would need to produce enough mountain lions so that dispersal would continue to replenish the sink areas. Dispersal can greatly improve the population projections for heavily hunted, or sink, populations.

There are likely to be different perspectives on which particular areas should be open or closed to hunting. To minimize conflicts we suggest a multiple stage decision process. First is for WGFD, working closely with the U.S. Forest Service, the National Park Service, and the public, to determine what areas might already be acting as reservoir populations by default and might be likely to continue to function as such. Some county and national forest lands are already closed to hunting. Although county commissioners have been working hard to protect wildlife areas, there are limitations to what they can do. Bridger-Teton National Forest has also made strides to protect larger tracts of winter range for deer and elk.

The agencies should determine if existing winter closure areas provide adequate habitat for mountain lions and if they provide good corridors into sink areas. WGFD believes at present that even without additional winter closures there are enough closed areas in Grand Teton National Park, the National Elk Refuge, and some partially closed wilderness areas to protect mountain lions.[49] This remains unsubstantiated. Full closures account for only 902 square miles out of 4,916 square miles in Jackson Hunt Area 2.[50] These areas do not contain a large year-round population of mountain lions.[51] To establish additional source and sink zones, both biological and sociopolitical contextual factors will need to be examined.[52] Habitat quality and lion numbers will need to be analyzed. Estimates of the number of resident animals in each protected area can then be calculated, representing what the state currently has as a baseline number of protected animals. One option might be to designate all of Teton County as a source area, creating a buffer for the national parks and a source population for the rest of the state, and establishing a policy that matches the views of the people in the region. Problematic animals can be removed on a case-by-case basis. The next step is to determine "buffer" areas that may be more accessible but still remote enough that they receive little hunting pressure.[53]

All this information would give managers some idea of which subpopulations are currently fully or partially protected. Population densities for each area can be derived either from field efforts or (at least initially) estimates based on current literature. The sum of these estimates will give managers and citizens an idea of how many mountain lions are currently protected, and the cooperative plotting of these areas on a map will show their proximity to one another and allow estimates of lion numbers protected on a regional basis. Disagreement is to be expected, and the process will require time, patience, diplomacy, reliable data, and possibly conflict management, including facilitation or mediation.

With this baseline information, managers and citizens can begin the next step of identifying population gaps (insufficient numbers of lions being protected in a region) or area gaps (inadequate configuration of subpopulations to allow the metapopulation dynamics to operate). In some regions sufficient mountain lion numbers and adequate configurations of protected areas already exist, and management of lion hunting will require only formalizing the setups that already exist. For other areas, decisions will need to be made regarding which areas to close and which to leave open for hunting. Some areas will require only a few modifications, such as closing some hunting units to provide corridors for the metapopulation dynamics to operate or to boost subpopulations in reservoir areas.

Using the metapopulation approach or not, the hunting quotas post-1999 may be too high to maintain viable populations in the long term. Although there is currently a dearth of population data, empirical data from other areas and population modeling can be used to project populations. High adult mortality, particularly of females with cubs, can cause mountain lion populations to decline significantly. Nevertheless, a WGFD official stated that the state had experimented in Star Valley by hunting mountain lions until the population was cut in half, and yet the population still rebounded.[54] The goal of any wildlife management process should be to maintain healthy, viable populations over time. Although modifications can be made on a year-to-year basis in case of drought or disease, in general, the quotas should be based on long-term goals.

Management based on a metapopulation approach should satisfy both hunters and conservationists and reduce conflict among opponents since it allows hunting while at the same time ensures long-term viability of mountain lions based on biologically sound methods. There are two key advantages to such an approach. First, given reasonable minimum and maximum estimates of resident lion densities, sufficient numbers of subpopulations can be maintained free of hunting pressure, thus ensuring long-term survival of the regional (or meta-) population, regardless of the hunting pressure exerted in the open areas. Second, since the designation of the amount of protected areas is based on the historic minimum density of resident animals, this eliminates the need for annual estimates of mountain lion densities.

Conclusion

Mountain lion management can be difficult, as the situation in western Wyoming so well illustrates. WGFD's management decision process has become politicized because of complex, interactive factors that reflect the values at stake and the nature of the institutions of decision making. Three of the primary issues raised by the current management difficulties are distinguishing and identifying the values that people have at stake, clarifying the goals of management, and including a role for the public in management. We recommend that WGFD create a formal management plan based on meaningful public input and broad-based public support. People need to be able to understand the basis of the agency's lion management policy. As well, the agency's decisions need to be open, fair, factual, timely, and justified. Improving public education and protecting wildlife habitat and corridors can have a significant influence on mountain lion protection, particularly in Teton County. Finally, a metapopulation approach to mountain lion hunting that establishes unhunted subpopulations and sets reasonable hunting quotas, based on both empirical data and population modeling, can secure viable mountain lion populations for the future. This combination of actions can help reduce the politicization that has dominated mountain lion management in recent years and help conserve Wyoming's cultural and natural heritage.

References

1. M. Hansen, 1999, "Public weighs state plan to increase cougar hunting," *Jackson Hole Guide,* June 30, A8; C. Blessley, "Lions should be subject to better management," *Jackson Hole Guide,* April 26, 2000; Odell, 1999, "36-lion hunting season sought," *Jackson Hole News,* June 16, A1, 34.; B. Holz, 2000, "Lion hunt justified," *Jackson Hole News,* February 23, A4; R. Huntington, 2001, "Lion-hunting opponents attack Game & Fish ad," *Jackson Hole Guide,* June 20, A8; W. Royster, 2001, "Game and Fish hears lion policy criticism," *Jackson Hole News,* June 20, A17; T. Mangelsen, 2001, "Game and Fish ignores mountain lion science," *Jackson Hole News,* July 18, A14.

2. The data on lion natural history came from the following sources: K. M. Murphy, P. I. Ross, and M. G. Hornocker, 1991, "The ecology of anthropogenic influences on cougars," 77–102 in T. W.Clark et al., *Carnivores in Ecosystems: The Yellowstone Experience,* Yale University Press, New Haven; W. B. Ballard et al., 2001, "Deer-predator relationships: A review of recent North American studies with emphasis on mule and black-tail deer," *Wildlife Society Bulletin* 29, 99–115; J. Laundré and T. W. Clark, 2003, "Managing puma hunting in the western United States through a metapopulation approach," *Animal Conservation* 6, 159–170; K. Logan and L. Sweanor, 2000, "Puma," in S. Demarais and P. Krausman, eds., *Ecology and management of large mammals in North*

America, Prentice Hall, Upper Saddle River, NJ; C. A. Lopez Gonzalez, 2001, "Implicaciones para la conservacion y el manejo de pumas (*Puma concolor*) utilizando como modelo una poblacion sujeta a caceria deportiva," Ph.D. dissertation, Universidad Nacional Autonoma de Mexico; J. Weaver, P. Paquet, and L. Ruggiero, 1996, "Resilience and conservation of large carnivores in the Rocky Mountains," *Conservation Biology* 10, 964–973. For research on mountain lions in Western Wyoming, see K.A. Logan, L.L. Irwin, and R. Skinner, 1986, "Characteristics of a hunted mountain lion population in Wyoming," *Journal of Wildlife Management,* 50, 648–654; and for ongoing research, see http://www.predatorconservation.org/predator_info/Forest_Clearinghouse/Cougar/cougarclear.htm; http://www.cougarfund.org/programs.html (scroll down for SCIENCE/SCIENTIFICRESEARCH), and http://www.beringiasouth.org/index.cfm=?id=tetonCougarProject.

3. Laundré and Clark, "Managing puma hunting in the western United States: Through a metapopulation approach."

4. Historical information was gathered from: Logan and Sweanor, "Puma"; WGFD, 1996, *Draft mountain lion management plan,* C. Anderson, ed., WGFD, Cheyenne; Murphy et al., "The ecology of anthropogenic influences on cougars"; WGFD, 1999, *Annual report of big game harvest,* WGFD, Cheyenne.

5. J. W. Laundré, L. Hernández, and T. W. Clark, in manuscript, "Population trends in pumas and mule deer and their possible causative factors, *Journal of Wildlife Management.*

6. Odell, "36-lion hunting season sought."

7. R. Huntington, 2003, "Commission locks in higher cougar quotas," *Jackson Hole News & Guide,* August 6, A9–10.

8. D. Bjornlie and D. Moody, 2003, *Annual mountain lion mortality summary, harvest year 2002,* Trophy Game Section, WGFD, Cheyenne.

9. Huntington, "Lion-hunting opponents attack Game and Fish ad."

10. See http://gf.state.wy.us for information on lion hunting quotas.

11. Michael Wolf (Utah State University), pers. comm., December 2000.

12. N. Breitenstine, 2002, "Feeding wild animals can make people prey," *Jackson Hole News,* January 16, A5.

13. *Casper Star Tribune,* October 6, 1999, B1; *Casper Star Tribune,* November 4, 2001, C1.

14. J. Deurbrouck and D. Miller, 2001, *Cat attacks: True stories and hard lessons from cougar country,* Sasquatch Books, Seattle.

15. S. Riley and D. Daniel, 2000, "Wildlife stakeholder acceptance capacity for cougars in Montana," *Wildlife Society Bulletin* 28, 931–939.

16. T. Mangelsen and C. Blessley, 2000, "Killing many cougars harms natural balance," *Jackson Hole News,* February 16, A5; C. Blessley, 2000, "Mountain lions need a Jackson Hole refuge," *Jackson Hole Guide,* April 26, A5; T. Mangelsen, 2000, "Out for blood: A look at state mountain lion management policies," *Alliance News* (Jackson Hole Conservation Alliance), Summer Issue; I. Rutkovskis, 2000, Letter to the edi-

tor: "Cougar compassion," *Jackson Hole News,* February, 23, A11, A22; K. Shea and R. Shea, 2000, "Where's the science?" *Jackson Hole Guide,* March 8, A4; T. Mangelsen, 2000, "How do you sleep?," *Jackson Hole News,* March 1, A24; A. Smith, 2000, "Impoverishment," *Jackson Hole News,* March 1, A25; M. Grant, 2000, "Privileged," *Jackson Hole News,* March 22, A24; Huntington, "Commission locks in higher cougar quotas"; T. C. Mayo, 2003, Letter to the editor: "Cougar call arrogant," *Jackson Hole News&Guide,* August 6, A19–20; R. Huntington, 2003, "Mountain lion quota may rise in hunt area," *Jackson Hole News&Guide,* June 4, A25; R. Huntington, 2003, "Public split over lion hunting quota," *Jackson Hole News&Guide,* June 11, A11; T. Mangelsen, 2003, Letter to the editor and to WGFD: "Revisit cougar quotas," *Jackson Hole News&Guide,* July 2, A21.

17. L. Dorsey, 2001, Letter (and e-mail) to John Baughman, director of WGFD, Cheyenne, November 9, signed by additional members of the conservation community.

18. WGFD, *Annual report of big game harvest.*

19. J. Turner, 2000, "Lion hunters display the courage of pooh," *Jackson Hole News,* February 23, A5; Holz, "Lion hunt justified"; Mangelsen, "Game and Fish ignores mountain lion science."

20. Rutkovskis, Letter to the editor: "Cougar compassion."

21. WGFD, *Draft mountain lion management plan.*

22. WGFD, *Draft mountain lion management plan.*

23. Anonymous WGFD staff member, pers. comm., March 2002.

24. WGFD, 2001, "Mountain lion management in the Jackson/Pinedale region," public announcement issued by Pinedale Regional Office, *Jackson Hole Guide,* June 13, A22, A25.

25. Bernie Holz, pers. comm., March 2002.

26. WGFD, "Mountain lion management in the Jackson/Pinedale region."

27. Mangelsen and Blessley, "Killing many cougars harms natural balance"; C. Blessley, 2000, "Mountain lions need a Jackson Hole refuge"; Mangelsen, "Out for blood: A look at state mountain lion management policies"; Huntington, "Commission locks in higher cougar quotas"; Mayo, Letter to the editor: "Cougar call arrogant"; Huntington, "Mountain lion quota may rise in hunt area"; Huntington, "Public split over lion hunting quota"; Mangelsen, Letter to the editor and to WGFD: "Revisit cougar quotas."

28. T. Mangelsen and C. Blessley, 2000, *Spirit of the Rockies: The mountain lions of Jackson Hole,* Images of Nature, Omaha.

29. Rutkovskis, Letter to the editor: "Cougar compassion"; Shea and Shea, "Where's the science?"; Mangelsen, "How do you sleep?"; Smith, "Impoverishment"; Grant, "Privileged"; Huntington, "Commission locks in higher cougar quotas"; Mayo, Letter to the editor: "Cougar call arrogant"; Huntington, "Mountain lion quota may rise in hunt area"; Huntington, "Public split over lion hunting quota"; Mangelsen, Letter to the editor and to WGFD: "Revisit cougar quotas."

30. R. Odell, 2000, "Area lion hunt hit," *Jackson Hole News,* April 26, A1, A31; Royster, "Game and Fish hears lion policy criticism."

31. T. Mangelsen, pers. comm., March 2002.

32. Turner, "Lion hunters display the courage of pooh."

33. Mangelsen, "Game and Fish ignores mountain lion science."

34. Anonymous WGFD staff member, pers. comm., March 2002.

35. M. P. Pimbert and J. N. Pretty, 1995, *Parks, people, and professionals: Putting "participation" into protected area management,* Discussion Paper DP 57, U.N. Research Institute for Social Development, Geneva, Switzerland.

36. R. Rundquist, 2001, "Insular agency," *Jackson Hole Guide,* July 11, A5, A25.

37. WGFD, "Mountain lion management in the Jackson/Pinedale region."

38. Huntington, "Commission locks in higher cougar quotas."

39. Huntington, "Commission locks in higher cougar quotas."

40. Huntington, "Commission locks in higher cougar quotas."

41. Huntington, "Commission locks in higher cougar quotas."

42. Mayo, Letter to the editor: "Cougar call arrogant."

43. See, for instance, http://www.buglebasinoutfitters.com, http://www.doublediamondoutfitters.com, http://www.windrivertrophyhunts.com, and http://www.JacksonHoleOutfitters.com (accessed 7/27/04).

44. T. W. Clark, 2002, *The policy process: A practical guide for natural resource professionals,* Yale University Press, New Haven.

45. S. Primm and T. W. Clark, 1996, "Making sense of the policy process for carnivore conservation," *Conservation Biology* 10, 1036–1045.

46. Laundré and Clark, "Managing puma hunting in the western United States through a metapopulation approach."

47. Odell, "Area lion hunt hit."

48. Laundré and Clark, "Managing puma hunting in the western United States through a metapopulation approach."

49. WGFD, "Mountain lion management in the Jackson/Pinedale region."

50. WGFD, "Mountain lion management in the Jackson/Pinedale region."

51. Odell, "Area lion hunt hit."

52. Laundré and Clark, "Managing puma hunting in the western United States through a metapopulation approach."

53. Laundré and Clark, "Managing puma hunting in the western United States through a metapopulation approach."

54. Anonymous WGFD official, pers. interview, March 2002.

Chapter **4**

Grizzly Bear Recovery: Living with Success?

Steve Primm and Karen Murray

Grizzly bears have long been a part of life and lore in the American West. Historical accounts, myths, and art celebrate grizzlies as revered warriors, fearsome adversaries, or the walking embodiment of the frontier.[1] Grizzly Basin, Silvertip Gulch, Grizzly Peak, and scores of other place names attest to the powerful symbolic hold of grizzlies on the imagination and identity of the West.[2] Having these rugged, independent creatures as neighbors is a point of pride for many people. Among other things, grizzlies may remind Westerners of the frontier era, indicating that they live in a minimally domesticated place, far from the safety and strictures of civilization.[3]

Such positive and inspiring symbolic images of bears may quickly fade, however, when a person has to deal with an actual grizzly killing cattle or tearing into a backcountry camp. Conflicts like these between grizzlies and people have fostered portrayals of bears in quite a different light: as dangerous killers that are unacceptable threats to the safety and livelihoods of hardworking American citizens. Rather than taking pride in the presence of bears, some people have responded by demanding that bears be eradicated from areas used by humans.

Unfortunately, conflicts between grizzly bears and humans have escalated over the past decade in the region covered by this book (western Wyoming south of Yellowstone National Park—see figure 1.1).[4] This increase is in part a reflection of the proliferation of grizzly bears throughout Greater Yellowstone, but, as we shall see, other parts of the ecosystem have not experienced such a dramatic increase in bear-human conflicts. It is unclear exactly why this region has a disproportionately high number of these conflicts. It is clear, though, that the rise in grizzly conflicts, in tandem with the emergence of wolf conflicts since 1995, has created significant and intransigent sociopolitical strife in the region. This sociopolitical climate is highly problematic for grizzly conservation, as well as intrinsically undesirable.[5]

In this chapter we examine the conservation implications of grizzly-human conflicts in western Wyoming. One might characterize grizzly bear recovery in Greater Yellowstone as "a victim of its own success," in that there are now enough bears to cause significant amounts of trouble. Recovery efforts have reestablished grizzlies not only in Yellowstone National Park, but more broadly in the Greater Yellowstone Ecosystem, a development that appears to have taken many individuals and institutions by surprise.[6] No one interested in successful wildlife conservation should be sanguine about the current level of grizzly-human trouble. On a biological level, unmitigated conflicts between bears and people will ultimately lead to bear deaths, either on a case-by-case basis or through the wholesale removal of grizzlies from vast areas. Social and political feedback from these superficially "biological" problems may do far more than merely halt further grizzly bear expansion in the West. Anticonservation politicians may use grizzly-human conflicts as ammunition to block unrelated initiatives or punish resource agencies by slashing budgets. In a worst case scenario, grizzly and wolf "horror stories" could be used to rationalize the evisceration of environmental laws such as the Endangered Species Act. On a more fundamental level, these conflicts, if not resolved in a fair and equitable manner, can lead to treacherous schisms within our society. These reasons provide sufficient justification for improving coexistence between grizzlies and people.

We begin the chapter with an overview of relevant natural history and ecological information about grizzly bears. Then we review the early management history of Greater Yellowstone grizzlies, highlighting factors that have led to the current level of conflict. This leads into a discussion of the increasing politicization of grizzly management over the last three decades. We argue that grizzlies have become a focal symbol in a larger cultural conflict throughout the American West. Pro- and anticonservation interests alike may be willing to manipulate grizzly-human conflicts to their own ends, rather than working together to mitigate or resolve conflicts. Improving practices for coexistence between people and bears is a critical task for conservation, but it is a task that is well beyond straightforward technical solutions. The chapter concludes, then, with practical recommendations for making progress in such a polarized context.

Natural History, Population Dynamics, and Early Management History

To understand grizzly-human conflicts, it is necessary to have some understanding of grizzly bears as a species. Fortunately, modern wildlife research has produced an impressive body of knowledge about these bears. We briefly review grizzlies' natural history, population dynamics, and early efforts to manage the relationship between them and us.

Natural History

Grizzly or brown bears belong to the single species *Ursus arctos*.[7] This species spans much of Earth's north temperate and arctic zones. *U. arctos* ranges from the Pyrenees Mountains of Spain eastward across Eurasia to the Wind River Range of western Wyoming. Across their range, grizzly/brown bears occupy boreal forests, arctic tundra, mountainous regions, and even some deserts. Their distribution is the widest of all bear species and among the greatest of all terrestrial mammals, reflecting the species' adaptability and omnivorous feeding habits.[8] Worldwide, *U. arctos* numbers approximately 200,000 individuals, with the bulk of these occurring in Russia.[9] Among the eight bear species of the world, grizzly/brown bears rank a distant second in numbers to the American black bear (*U. americanus*).[10]

Grizzly bears apparently evolved in Eurasia in the early to middle Pleistocene, or approximately 1–1.5 million years ago. Fossil records indicate that *U. arctos* migrated to North America between 50,000 and 70,000 years ago. Recent genetic analysis indicates that *U. arctos* in North America exists as four geographically separated genetic lineages, one of which is the Rocky Mountain grizzly bear. These lineages evidently diverged long before *U. arctos* migrated from Eurasia, suggesting that the different phylogeographic groups entered North America at different times. Geographic barriers such as major ice sheets could have minimized contact among these groups.[11]

Geneticists refer to these groups as "evolutionarily significant units," which are themselves a worthwhile focus of conservation efforts.[12] These genetically distinct groups represent both past and future adaptability within a species and may offer a rare opportunity to conserve evolutionary potential in a large carnivore. Adaptability allows a species to survive changing conditions, whereas evolutionary potential allows a species to evolve into a separate species over time.[13] Critics of Endangered Species Act protection for grizzlies in the contiguous United States argue that the relatively high numbers of *U. arctos* worldwide demonstrate that the species is not threatened or endangered. Geneticist Lisette Waits and colleagues counter that "genetic distinctiveness . . . highlights the importance of listing *U. arctos* populations in the lower 48 states [under the act] despite the fact that brown bears are thriving in Alaska."[14]

Grizzlies are intelligent, adaptive omnivores, capable of using diverse food sources. This adaptability allowed them to colonize habitats throughout the Rocky Mountains, eastward into the Great Plains, and westward through much of what is now California. Grizzlies feed on fish, ungulates, rodents, grasses, sedges, forbs, berries, mushrooms, insects, and mast crops such as acorns and pine seeds. Compared to wolves and big cats, grizzlies are relatively inefficient predators. Very young elk and moose calves are vulnerable to grizzly predation, as are mature male ungulates weakened by the breeding season. Grizzlies scavenge much of the meat they consume from carcasses of ungulates dead from other causes. In some habitats, grizzlies are largely vegetarian, even though they are unable to digest plant fiber fully. They are able to use plant foods by selectively consuming plants at the phenologic stage that offers the most nutrition.[15]

Grizzlies and brown bears have also demonstrated a marked propensity for exploiting foods associated with humans. Grizzlies readily adapt to eating crops, livestock, garbage, and other anthropogenic foods. Unfortunately, once a grizzly discovers high-quality foods associated with humans, lethal control of the bear is the most likely outcome. This is because bears are powerful, persistent, and willing to use aggression against competitors. Once exposed, there is a good possibility that grizzlies will damage property and potentially injure people in their efforts to acquire human foods.

Prior to the arrival of Euro-American settlers, who possessed firearms and other technologies, the consequences to grizzlies for these behaviors would have been relatively minor.[16] Once these settlers arrived, however, the full array of lethal technologies (e.g., traps, poisons, and firearms) was set against grizzlies. Depletion of natural food sources by humans and their livestock likely exacerbated conflicts with grizzlies and other predators. Because they are omnivorous, wide-ranging, large-bodied, and aggressive when threatened, grizzlies in the West were rapidly extirpated by Euro-American people.

The nature of contemporary conflicts and resultant bear mortalities indicates that North American grizzlies, in their two centuries of experience with technologically advanced people, have not evolved many behaviors that would foster coexistence. In recent decades, fortunately, people have managed to adopt certain behaviors that do allow some degree of coexistence. Preventing bears from getting human foods, providing bears with temporal and spatial refugia, and the development of nonlethal deterrents such as pepper spray all contribute to coexistence between people and bears.[17]

Population Dynamics

The high likelihood of conflicts between people and grizzlies makes conservation challenging. Compounding these difficulties, grizzly populations grow slowly, so they cannot recover quickly from declines. This attribute stems from inherently low reproductive rates. Female grizzlies may be 5 years old before giving birth and then may give birth only once every 3 years thereafter. Cubs stay with their mothers for 2 or 3 years. During this rearing time, the female must keep the cubs away from male grizzlies, who will kill the cubs given the opportunity, either as food or to eliminate a competitor's offspring. As a result, female grizzlies often end up in lower quality habitat (i.e., with poor food sources or in close proximity to people) to avoid encounters with mature males, who evidently have primacy over better habitat. The segment of the population that has the greatest effect on population growth may also be the segment that is most at risk for encounters with, and mortalities caused by, people.[18]

Thus, thriving grizzly populations require a good deal of habitat that is remote from people. Several studies have helped quantify what is required for secure habitat for grizzlies. Bear researcher David Mattson and colleagues found that during fall, when grizzlies are foraging intensively to develop fat reserves for denning, many bears did not come closer than 1.86 miles to primary roads and developments in Yellowstone National Park. Foraging patterns were disrupted at a distance of 2.5 miles from roads and developments.[19] Canadian scientists Bruce McLellan and David Shackleton found similar responses to roads in other areas, finding that "even a little traffic [on gravel, secondary roads] is sufficient to displace" grizzly bears from otherwise productive habitat.[20] These disruptive factors have had a major impact on female grizzlies (since they are more likely to have to forage near roads) and, consequently, negative impacts on reproductive rates.

Grizzlies that do not avoid roads and developed areas run into different problems. Bears that are unafraid of being near people and infrastructure are considered habituated, meaning they exhibit a neutral response—neither fleeing nor approaching—these stimuli. Habituated grizzlies can be dangerous to people, especially in a national park setting with large numbers of inexperienced, unarmed people. One of the main risks of habituated bears is that they will take the next step and begin to perceive that people or developed areas are sources of food. These "food-conditioned" bears are extremely dangerous because they are even bolder around people than habituated bears and may resort to aggression to obtain and defend anthropogenic foods.[21] Because they routinely end up in less secure habitats, female grizzlies with cubs are more likely than other bears to end up food-conditioned and thus are more likely to be killed or otherwise removed from the population. Roads and developments thus tend to render nearby habitat ineffective by causing bears to abandon it or by facilitating habituation and food conditioning, which may cause bears to be killed or relocated.[22] Although the Greater Yellowstone Ecosystem comprises some of the largest roadless areas in the contiguous United States, it still contains many roads and developments. Fifty percent of Yellowstone National Park, the ecosystem's protected core, is within 5 miles of a paved road and within 8 miles of a major development or village. Much of the region's occupied grizzly habitat is open to road building and other development activities.[23]

Even areas without roads pose challenges for grizzly conservation. Greater Yellowstone's extensive wilderness areas support a thriving big-game outfitting industry. Every fall, hunters, guides, and outfitters move into these areas in significant numbers to hunt elk and other big game.

Fall is also the time when grizzlies are busy consuming calories for their winter sleep. The hunting parties are associated with numerous attractants, including foodstuffs, livestock feed, and big game carcasses. Moreover, these people are armed and are moving quietly in search of game. In recent years, this juxtaposition of armed, stealthy people, significant attractants, and hungry bears has resulted in a number of violent encounters that have ended with dead grizzlies.[24] The dead grizzlies are often females with cubs, since they tend to be the most food-stressed and also more inclined to defensive aggression against perceived threats.[25]

Conservation of grizzlies in a landscape dominated by people, then, can be quite challenging. In the next sections, we review the trajectory of grizzly conservation in western Wyoming over roughly the last century.

Early Management History

Grizzlies once ranged throughout Wyoming, even seasonally occupying the state's arid Red Desert region.[26] By the 1920s, though, grizzly range had contracted severely throughout the West. The bears were first extirpated from open country and persisted in rugged mountain ranges.[27] Western Wyoming in particular served as a stronghold for grizzlies, because of its vast expanses of mountainous terrain with a large protected area, Yellowstone, as core habitat. Through the middle decades of the 20th century, grizzlies survived in this region while they became extinct throughout most of the West.[28]

In addition to its rugged terrain and low numbers of resident people, Yellowstone functioned as a grizzly refuge in two other ways. First, Yellowstone began protecting bears from unregulated killing in 1886, whereas killing outside park boundaries went unchecked until the 1970s. Second, from the park's earliest days, grizzlies and black bears had unfettered access to open garbage dumps within the park.[29] Hotels, administrative sites, and work camps all provided a steady and abundant supply of food scraps and other edibles. These reliable, concentrated sources of high-quality food tended to hold aggregations of bears inside Yellowstone, where they were safe from hunters, poachers, and stockmen. Also, the garbage dumps served as a source of entertainment for park visitors, who would gather on bleachers to watch bears congregate and eat garbage.[30] The grizzly population in

Yellowstone apparently remained stable up to the 1960s. Park historian Paul Schullery summarized annual population counts from 1920 to 1970; censuses ranged from 75 to 335 grizzlies during that period, with most counts falling between 200 and 300.[31]

There were evidently numerous grizzlies in much of the rest of western Wyoming for the first half of the century as well. For example, biologist Adolph Murie reported frequent predation on cattle by grizzlies on the Blackrock-Spread Creek grazing allotment northeast of Jackson Hole on the Teton National Forest.[32] Biologist Sanjay Pyare and colleagues reported that grizzlies persisted even in the southern extremities of the Greater Yellowstone Ecosystem through the 1930s.[33] By the late 1950s, however, many observers were concerned that grizzlies were on their way out. In 1956 Yellowstone's chief naturalist David Condin warned, "The future for this species is dark unless attitudes, policies and regulations change so the animal can re-establish itself on some former range areas and in addition, unless it is afforded greater protection in those ranges where it is now found."[34]

George Reynolds estimated in 1959 that only 25–50 grizzlies remained in Wyoming outside Yellowstone. Yet they were still being actively hunted. At the time, Wyoming elk hunting licenses included an entitlement to kill grizzlies. In effect, Wyoming was issuing 85,000 grizzly hunting permits annually.[35] Although some hunters voluntarily reported that they had killed grizzlies, there was no mandatory reporting until 1970, so total legal take was unknown.[36] Hunting regulations in Montana and Idaho were similarly liberal. Additionally, the Greater Yellowstone Ecosystem's grizzlies were freely killed in response to predation on livestock, especially domestic sheep. The extent of indiscriminate control kills (that is, killing grizzlies on sight or with poison baits without targeting a known problem bear) by stockmen is undocumented but was surely a factor in the population decline. Grizzlies and black bears were also routinely killed inside Yellowstone because of conflicts with visitors, such as injuries to tourists or damage to property. The park's control kills, however, were more selective.[37]

Thus, by the 1960s the grizzly bear population had largely contracted to the park and the immediately surrounding mountainous country. With the garbage dumps as a reliable, high-quality food source, Yellowstone's "carrying capacity" was artificially high, which kept the bear population concentrated safely within the park. Much of the area covered by this book was devoid of grizzlies.[38] Yellowstone National Park provided enough high-quality food—in the form of garbage—to maintain a stable source population of grizzlies, some of which would occasionally disperse into areas outside park boundaries. This dispersal would likely account for the continued presence of grizzlies outside the park in spite of low survival rates in such areas.[39]

Politicization of Management

As the 1960s arrived, grizzly populations in Yellowstone were apparently fairly stable, although concerns were mounting. Before the decade was out, however, major events would place grizzlies at the center of a national political controversy. Paradoxically, this era yielded vast new knowledge about grizzlies and increased public awareness, yet also pushed them near the brink of extirpation from the region.

Yellowstone Dump Controversy

In 1959 wildlife biologists Frank and John Craighead initiated a pioneering study of the grizzlies of Yellowstone National Park. This study ran until 1970 and focused heavily on grizzlies using garbage dumps in the middle of Yellowstone, though the Craigheads also monitored bears far from the dumps. From their groundbreaking use of radio telemetry to improved methods of immobilizing bears for research and management purposes, the Craighead study made monumental contributions to grizzly conservation and wildlife science. The study became globally known through articles in popular magazines and National Geographic films.[40]

Concurrently, the National Park Service embarked on an ambitious program to move the national parks toward more natural conditions. In Yellowstone the new policy meant that garbage would be made unavailable to bears. In addition to the move toward a more "primitive" state, the National Park Service was also responding to two grizzly-caused human fatalities in Glacier National Park on a single night in 1967. Both grizzlies implicated in the Glacier Park fatalities routinely ate garbage and other anthropogenic foods. Yellowstone began reducing the amount of garbage at the largest dump in 1968 and completely closed the dumps by 1971.[41]

Cutting off the garbage was controversial. The Craigheads openly disagreed with the Park Service over how to implement the closures, maintaining that bears should be slowly weaned from garbage over a 10-year period with careful monitoring of the responses of marked animals.[42] The National Park Service chose a much faster dump closure schedule and failed to monitor bear responses. According to the Craigheads, the National Park Service not only failed to monitor bears themselves, but also impeded the Craigheads' research efforts by removing artificial markings (e.g., ear tags) from grizzlies, inadequately documenting management actions and lethally controlled bears, and eventually bulldozing a Yellowstone building that the Craigheads had used as their field headquarters.[43] The Craighead study, though tremendously productive and world renowned, ended in acrimony. The Craigheads severed their relationship with Yellowstone National Park in 1970 and terminated their study.

The Park Service's decision to phase out the dumps rapidly initiated a period of unsustainably high grizzly mortality in Yellowstone and surrounding areas. Figure 4.1 illustrates the mortality that occurred from 1959 to 1975, when grizzlies were protected under the 1973 Endangered Species Act. The pronounced mortality spike in 1967 (40 grizzlies killed) was largely the result of legal hunting kills in Wyoming; apparently the Wyoming Game & Fish Department's (WGFD) announcement that the grizzly season would be closed the following year (1968) led to this upsurge.[44] As noted above, Wyoming hunters were not required to report killed grizzlies until 1970, so the legal take in Wyoming may have been higher than shown. Wyoming resumed its grizzly bear hunt in 1970 on a far more limited basis.[45]

Recorded control kills inside Yellowstone increased rapidly from 1968, when the reductions in garbage at the dumps began, and reached a peak in 1970. That year 23 grizzlies were killed inside the park and 31 in the states surrounding it. With no formal research and monitoring program after the Craigheads departed, these numbers are debatable. Thereafter, mortality appears to have shifted outside the park, perhaps because grizzlies were dispersing in search of new food sources. These grizzlies may have been unaccustomed to foraging for natural foods and were thus highly prone to trouble with people as they sought garbage, camp foods, and livestock. For example, the slight rise in Idaho conflicts in 1971–1972 was largely the result of grizzlies killed for preying on domestic sheep.[46] By 1973 it is likely that most

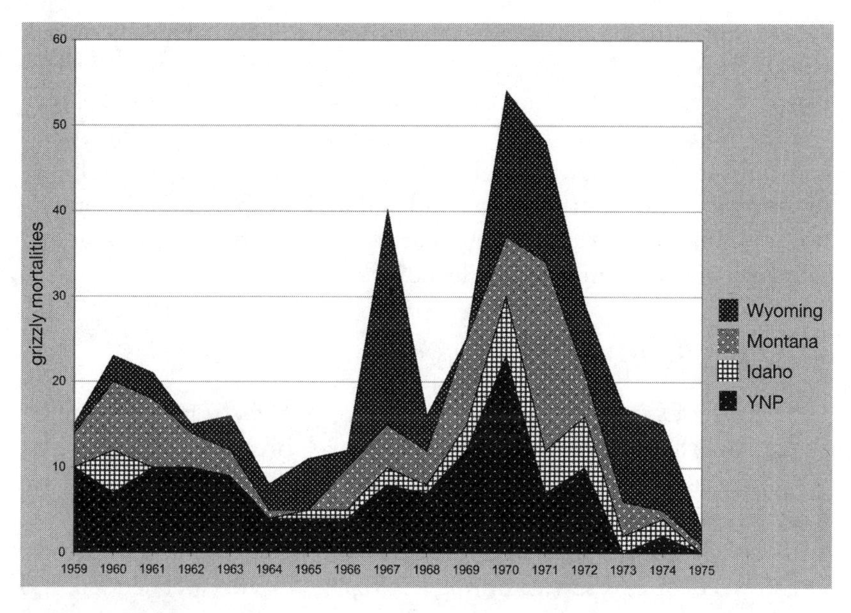

Figure 4.1
Human-caused grizzly bear mortalities in the Greater Yellowstone region by jurisdiction, 1959–1975 (data source: Craighead et al. 1988). YNP = Yellowstone National Park.

potential "problem" grizzlies were already removed from Yellowstone. The previous 4 years (1969–1972) saw a total of 156 recorded grizzly mortalities. The continued overall decline in mortality thereafter may reflect improved management practices, but it may also reflect the fact that the Greater Yellowstone Ecosystem was simply running out of grizzlies.

This catastrophic period was marked by public disputes between the park and the Craigheads and may partially explain the level of distrust and conflict that marks grizzly conservation today. The agencies responded to the crisis by creating the Interagency Grizzly Bear Study Team in 1973 to pick up the research and monitoring that had lapsed when the Craigheads left Yellowstone. The agencies also established the Grizzly Bear Steering Committee, a management counterpart to the Study Team.[47] Concurrently, the U.S. Department of the Interior (which oversees the National Park Service and the U.S. Fish and Wildlife Service) responded to the ongoing controversy by commissioning a

National Academy of Sciences committee to study the Yellowstone grizzly situation. This panel, the Committee on Yellowstone Grizzlies, issued its final report in July 1974. The report concluded that Yellowstone's policies had initiated a period of high grizzly bear mortality and criticized the park for failing to maintain adequate research and monitoring. The report also concluded that the grizzly population was not in "immediate danger of extinction," but strongly recommended intensive efforts to minimize mortality.[48]

Protection under the Endangered Species Act

In 1975 the U.S. Fish and Wildlife Service (USFWS) protected grizzlies under the Endangered Species Act as a "threatened" species. The act directs agencies "to use . . . all methods and procedures which are necessary to bring any endangered species or threatened species to the point at which the measures provided pursuant to this Act are no longer necessary."[49] Sport hunting of grizzlies immediately ended, and land management agencies began efforts to protect grizzly habitat and reduce conflicts.

Listing grizzlies under the Endangered Species Act was extremely controversial. Stockmen claimed that if grizzly numbers expanded and they were not allowed to kill problem individuals, ranching would become unprofitable. Hunters and outfitters did not want to lose valued hunting opportunities. Other interests were alarmed at the prospect of designating "critical habitat" for grizzlies—estimated by some to be around 20,000 square miles. The USFWS held more public hearings for grizzly bear critical habitat designation than for any other species. In late 1976 the U.S. Senate even held a special hearing in Wyoming—the only legislative review of a proposed critical habitat designation prior to the hearing for the northern spotted owl in 1991.[50] In the end, the USFWS did not formally designate critical habitat for grizzlies. Instead, the agency adopted a stratified zoning system that established grizzly conservation as the top priority in some areas and gave progressively higher priority to other goals in other places. The areas where grizzlies receive top priority became known as the Recovery Zone. Designating these areas was the responsibility of the relevant land management agency, in most cases the U.S. Forest Service (USFS). In some places the Recovery Zone boundary follows existing wilderness boundaries.[51]

With Endangered Species Act protection, an acute crisis for Yellowstone grizzlies had apparently ended, although an appreciable rebound from this close brush with extirpation would be more than a decade away. The population had dropped precipitously from a Craighead-era estimate of approximately 300 to fewer than 200.[52] Despite the act's shield, the grizzly population remained in dire trouble through the 1970s. "Garbage bears" remaining from the open-dump era and offspring that probably learned such behavior from their mothers continued with their old behaviors.[53] People in the area had not adopted behaviors that would reduce bear mortality and promote coexistence; thus, grizzlies continued to die in unsustainable numbers. In addition, fecundity may have declined as female bears went through a period of low nutrition while they learned to depend on natural foods.[54]

A leaked National Park Service memo in 1982 raised the alarm about the continuing precarious state of grizzlies. Park Service biologist Roland Wauer's memo detailed recent bear mortalities, noting that adult females had dropped to approximately 30 and that the total population was less than 200 and dropping. He asserted that "we no longer have the luxury of time to research the remaining parts of the puzzle. The Yellowstone grizzly bear picture is presently sufficiently clear enough! . . . It is imperative that highest priority be given to eliminating grizzly bear mortality."[55] The memo attracted national attention. Richard Knight, leader of the Interagency Grizzly Bear Study Team, told the *New York Times,* "We're looking at the end of the grizzly bear population in 20 to 30 years . . . something drastic needs to be done."[56] Also in 1982 the USFWS released the first version of its Grizzly Bear Recovery Plan. Generally, the plan identified causes of the decline in grizzly numbers, factors limiting population growth, and actions necessary to reach recovery goals. The plan expressed recovery goals in terms of grizzly bear natality and mortality, calling for extensive and expensive monitoring efforts. These goals were revised with the 1993 Recovery Plan, which relied on parameters that were more easily estimated.[57]

In 1983 the agencies revamped their grizzly conservation efforts with a new coordinating body called the Interagency Grizzly Bear Committee (IGBC), including representatives from the National Park Service, USFS, USFWS, and state wildlife agencies. This committee replaced and greatly expanded the old Grizzly Bear Steering Committee. The IGBC later expanded to include representatives from the Bureau of Land Management, the Bureau of Indian Affairs, and the provinces of British Columbia and Alberta.[58] The creation of the IGBC is noteworthy in its recognition that grizzly conservation and management is inherently a multijurisdictional task. The IGBC coordinates research and management of grizzlies in the contiguous United States. Since 1983 the IGBC has made significant progress in reducing grizzly mortality. This effort includes programs to prevent bears from accessing garbage as well as regulations mandating proper storage of human foods in the backcountry.[59]

Meanwhile, throughout the 1980s grizzlies were learning to rely almost exclusively on natural foods.[60] Lacking any single concentrated major food source such as salmon runs, a return to natural foraging may have dispersed the Yellowstone grizzly population throughout remote wilderness areas. Moreover, since they were no longer seeking foods associated with humans, encounters and conflicts between people and grizzlies declined in comparison to the 1960s and 1970s.[61] These developments also coincided with a period of high productivity for key natural foods, especially whitebark pine seeds.[62]

Whitebark pine seeds are a very important food for Yellowstone grizzlies. The seeds are high in fat, which is important for bears going into hibernation. Consumption of whitebark seeds is also associated with an increased likelihood of a female giving birth to cubs and with an increased likelihood of a large litter.[63] Furthermore, whitebark pine grows at higher elevations (above 8,000 feet); when bears are eating whitebark seeds, they are obtaining a high-quality food in a place where they are unlikely to encounter people.[64]

Thus, a series of years with abundant whitebark seed crops, along with intensive efforts to minimize grizzly mortality, combined to help the population of bears rebound during the 1980s and early 1990s.[65] In 1988 major forest fires burned through the region, including significant portions of Yellowstone's grizzly habitat. Although the 1988 fires eventually led to increased production of certain bear plant foods by opening up forested areas, in the short term the fires apparently reduced abundance of many bear foods. As a result, grizzlies may have shifted distribution away from these burned areas in search of more productive food sources.[66] Combined with an increased population size, this shift in distribution set the stage for some surprising developments in western Wyoming.

Range Expansion and Conflicts in the 1990s

In 1992 cowboy Terry Schramm began discovering dead cattle on the Blackrock-Spread Creek cattle allotment, the same area Adolph Murie studied in the 1940s. Blackrock-Spread Creek abuts the southern edge of the vast Teton Wilderness Area, which is contiguous with the remote backcountry of Yellowstone. Much of this 137-square-mile USFS grazing allotment lies inside the designated Grizzly Bear Recovery Zone under the USFS Recovery Plan. Schramm worked as a range rider for Jackson, Wyoming, rancher Paul Walton. Jackson Hole newspapers carried photographs of dead, mostly consumed calves. Schramm noted that "in the 13 years that I've been up there, I've never had a problem like this [grizzly predation] before."[67] Grizzlies had long been present on the allotment, but the extent of the conflicts was apparently a new development.[68]

Schramm documented seven grizzly-killed cattle in 1992, although other cattle went missing without any evidence of their fate. In accordance with state law, WGFD compensated the Walton Ranch for the documented losses.[69] Compensation for the projected market value of dead livestock is important but not satisfactory to many ranchers.[70] Predation by large carnivores carries other costs to ranchers. Ranch personnel must spend time seeking missing livestock and then additional time to document predator kills adequately to receive compensation. These activities take time away from other responsibilities, such as providing veterinary care to sick or injured livestock, maintaining fences, or any number of routine chores. Predation episodes also tend to panic livestock. The stress and the effort expended to flee predators causes livestock to lose weight and perhaps become susceptible to disease. Also, panicked livestock are more difficult to keep in designated grazing areas, which can lead to range damage if stock concentrate in certain areas to avoid predators. Schramm summed it up in the vernacular, "The difference between a bear walking through a bunch of cattle and one running out there and grabbing one and having it go squalling and bawling as it's dying is a whole different ballgame."[71] The problem temporarily went away in fall 1992 when the ranchers moved the cattle back to winter pastures and the bears went into their dens.

In July 1993 the cattle and the bears were back on the Blackrock-Spread Creek allotment with the same result. Local newspapers ran headlines like "Rogue grizzly bear attacks again."[72] By the end of the grazing season, the ranchers were seeking compensation for 46 calves, and at least ten more were unaccounted for. Paul Walton expressed his frustration to the *Jackson Hole News,* "We could have killed a dozen bears. We haven't even shot at one. We've been just squeaky clean. For the sake of the bears, they ought to do something. Grizzlies will run into other ranchers' stock, and these other people won't be like me."[73]

Schramm expressed similar sentiments, "I'm just a cowboy and I just started out to try and get the government agencies to help me out, and it's turned into a public lands issue. Now we have to decide whether we want to turn this into a grizzly bear park or we want to deal with the bears on a problem basis. My conflicts aren't going to be the last conflicts."[74]

Schramm's comments allude to a central point in the Blackrock-Spread Creek situation. Because the allotment was partly inside the Recovery Zone, pro-grizzly conservation organizations felt they had to "hold the line" and assert the needs of grizzly bears over livestock.[75] But because this was a long-existing grazing allotment, pro-livestock groups saw the situation as a case of callous bureaucrats and unfeeling environmentalists refusing to allow ranchers to protect their livestock. Those who favored delisting grizzlies cited the emergence of these conflicts as evidence that grizzlies were doing fine and needed to be controlled.[76] The case had become an important symbol in the broader fight over public land management.

The early- to mid-1990s were particularly tense years, as then-President Bill Clinton's Interior Secretary Bruce Babbitt attempted sweeping changes in public land use policies in the West. These reforms included reintroduction of wolves and changes in grazing policy and mining law. The Washakie County, Wyoming, board of commissioners represented the sentiments of many when they implored President Clinton to fire Babbitt, saying that his policies were "nothing less than bigotry carried out under Color of Law . . . that would decimate the custom and culture of the West through radical and unreasonable water and land-use policy change."[77] In this climate, it is easy to see how a story involving large, dangerous predators like grizzlies, heroic hardworking cowboys, and mutilated Hereford calves could become a focal point for various frustrations.

Schramm and Walton also proved to be prophetic. After 1993 grizzly-human conflicts began to escalate. Figure 4.2 shows the dra-

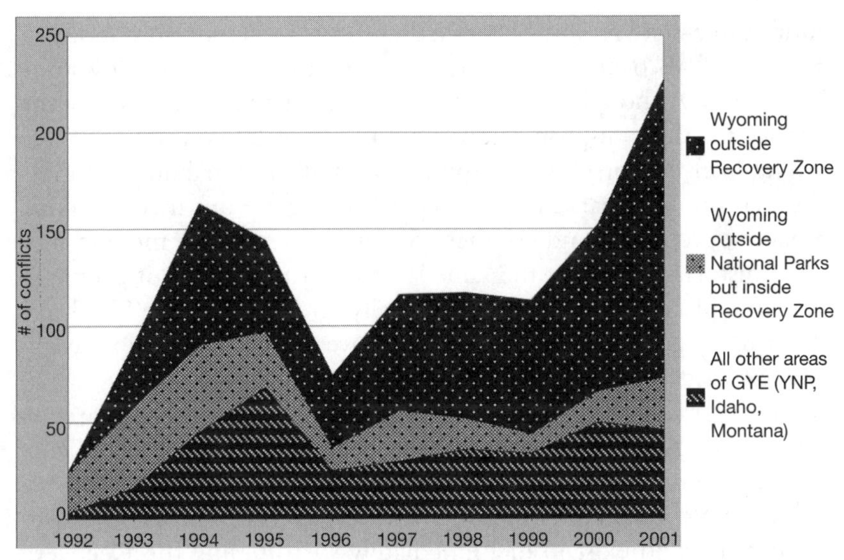

Figure 4.2
Grizzly-human conflicts in the Greater Yellowstone region, 1992–2001 (data source: IGBC annual conflict summaries, 1993–2001). YNP = Yellowstone National Park.

matic increase in grizzly conflicts (including livestock predation, bears damaging other property, and bears accessing human foods) in parts of Wyoming outside the Recovery Zone from 1992 to 2001.[78]

As the graph indicates, there were virtually no grizzly-human conflicts outside the Recovery Zone in Wyoming in 1992 (the first year of unified record keeping). The numbers began to climb from there. The general decline in 1996 is attributable largely to a bumper crop of whitebark pine seeds throughout the region. After 1996 conflicts outside the Recovery Zone began to increase again, while conflicts elsewhere leveled off. The decline in conflicts in areas within the Recovery Zone is largely the result of the translocation (in 1995) and then killing (in 1996) of a single stock-killing grizzly that had been active on the Blackrock-Spread Creek allotment. In 1999 ranchers vacated Blackrock-Spread Creek, which further reduced conflicts inside the Recovery Zone.[79] In 2003 a coalition of conservation organizations purchased the rights to the Blackrock-Spread Creek grazing permits, permanently retiring the allotment from livestock use.[80]

From 1997 through 2001, areas of Wyoming outside the Recovery Zone accounted for 60 percent of all grizzly-human conflicts in the entire three-state region. Spatially, these areas represent only about 17 percent of all occupied grizzly range in the Greater Yellowstone Ecosystem. Although it is impossible to determine conclusively the reasons for these disproportionately high levels of conflict, there are several likely factors. First, a pronounced shift in distribution of grizzlies has put the bears in places where people are not used to taking proactive steps to avoid conflicts. Second, grizzlies are moving into these historic habitats at the same time that the number of people in the same places is increasing dramatically. Areas such as the Shoshone River drainage near Cody, for instance, have experienced major booms in rural residential development in the past 15 years.[81]

Third, grizzlies are now occupying areas of relatively intense livestock grazing, including domestic sheep grazing.[82] Much of the core Recovery Zone has no livestock grazing, since it is either within Yellowstone National Park or remote wilderness areas. The livestock operations that do exist in or immediately surrounding the Recovery Zone are mainly cattle operations (there is one domestic sheep operation inside the Recovery Zone, near Gardiner, Montana; the rest have been phased out). Grizzlies and other large predators cause mayhem in domestic sheep bands, as the sheep panic and excite predators to even more killing. Sheep individually are easy for predators to kill and may end up dying en masse by stampeding off cliffs, into bodies of water, or into obstructed areas where they pile onto one another and suffocate. In short, many of the new conflicts are the result of grizzlies and other large predators expanding outward from the Recovery Zone and encountering sheep. Cattle predation continues to be a problem as well.[83]

In July 1999 the USFWS approved new guidelines for dealing with stock-killing grizzlies, particularly in the Upper Green River drainage near Pinedale, Wyoming. The guidelines allow stock-killing grizzlies to be killed if nonlethal measures fail and predation on livestock persists. Shortly after the USFWS issued the new guidelines, a large male grizzly was trapped and euthanized for killing calves on the Upper Green grazing allotment. In response, eight regional conservation organizations demanded that WGFD and the USFWS rescind the new guidelines and take a more lenient approach to stock-killing bears. Stepped up management efforts—such as removal of livestock carcasses to reduce attractants and increased human presence—and the removal of the known stock killer may have contributed to a temporary decline in livestock incidents.[84] Through 2003, however, grizzlies continued to kill cattle in the Upper Green.[85]

It is clear, then, that there has been a major change in the relationship between people and grizzlies in the past decade. For many years, there were few grizzlies at all outside Yellowstone National Park and its immediate surroundings. Most grizzlies were deep inside the areas that became the Recovery Zone. When the Recovery Zone was declared, it established expectations about where bears would exist and, more importantly, where they would have priority. The significance of these expectations was evident in the stance of environmentalists concerning the conflicts at Blackrock-Spread Creek. They felt that they had to "hold the line" because to do anything other than advocate for bears over cattle would erode expectations about grizzlies inside the Recovery Zone, with serious consequences for the bears.[86]

A clear set of expectations may be important, but the Recovery Zone also may have fostered a rigidly dichotomous way of looking at grizzly habitat. That is, people began to expect grizzlies inside the Recovery Zone and expect them not to be outside it. If grizzlies were outside the Recovery Zone, the expectation might be that they do not need to be there (i.e., if they have filled up the Recovery Zone, they must be recovered and doing fine) and that they should be removed (translocated or killed) at the first sign of trouble. Combine these violated expectations with the symbolism of grizzlies and perceptions of a "War on the West," and we have a formula for high levels of sociopolitical strife.

The controversy over grizzlies in western Wyoming is, as we have seen, partly an on-the-ground, ecological phenomenon that has been taking shape for nearly 40 years. Grizzlies really are in more places than they used to be, and they are in places where people were not prepared to deal with them. A conventional approach to such problems would treat this as a matter of designing and applying technical solutions. For instance, the problem of keeping grizzlies out of anthropogenic foods in the backcountry may be easily solved with proper bear-resistant containers and other equipment. Although such measures are necessary, the polarized context of grizzly expansion in Wyoming makes them insufficient. The problems we face are far more than technical problems; we face a conflict of values and must use tools that are appropriate for this kind of problem. In the next section we examine this polarized context in more detail and draw the connections between disputes over grizzly conservation and larger conflicts about the future of the American West.

Polarized Sociopolitical Context

It is difficult to ascertain the number of people whose day-to-day lives have been affected by grizzly expansion into long-vacant habitat. The list would include several ranching families, their employees, some backcountry outfitters, and a few dozen agency employees. An unknown number of recreationists and hunters now need to take precautions while afield, but their activities are seasonal and not related to making a living. Very rarely, someone is injured or killed by a grizzly bear. No one has been killed by a grizzly since 1986, although some people have been gravely injured since then. However, the odds of being injured by a grizzly are quite low, as data from Yellowstone National Park indicate: "From 1980–2002, over 62 million people visited [the park]. During the same period, thirty-two people were injured by bears. The chance of being injured by a bear while in the park is approximately 1 in 1.9 million."[87]

Nonetheless, one might get the impression that hundreds if not thousands of people are very upset about the return of grizzlies to their area. Three counties in western Wyoming—Fremont, Sublette, and Lincoln—have passed resolutions opposing grizzlies and wolves, along with legally impotent ordinances banning the big predators from their counties.[88] The ordinances are unenforceable because states, not counties, have jurisdiction over wildlife, and the federal government has jurisdiction over species protected by the Endangered Species Act.

It is not that the counties were unaware of the limits of their jurisdiction. The counties' resolutions and ordinances were clearly symbolic, as Lincoln County Commissioner Stan Cooper told the IGBC: "We're not a bunch of renegades and rednecks. We're just trying to get your attention."[89] These actions were apparently triggered by USFS plans to expand regulations requiring proper storage of anthropogenic foods into national forest areas far south of the Recovery Zone. The regulatory expansion was part of an agencywide effort to reduce human conflicts with both grizzlies and black bears, in part because of concerns about litigation should a known problem bear harm a person. Although it is apparent that the USFS made several strategic blunders in attempting these changes—for example, by expanding the regulations to a very large area in a short time, rather than phasing in the changes—it is equally apparent that certain factions in the three counties exploited the effort to foment hysteria about grizzlies.

These political factions found fertile ground for cultivating anger. Grizzlies are a powerful and often negative symbol for many people. These people see the protection of bears as a surrogate being used by conservationists and others to accomplish a much larger and more sinister agenda, as explained by Fremont County rancher and historian Lois Herbst: "It's [the larger agenda] called the Wildlands Project, Wildlands Recovery, or Wild Again. . . . They give it different names. It speaks about using the grizzly bears, the wolves and the lynx—the Canadian lynx—to end multiple use on the public lands. They want to make these lands wild again, and the states that have so much public land are the ones being hit the hardest, because they know they can get rid of us."[90]

The increase in conflicts provides people with more evidence of this supposed scheme, since it seems like one of the few plausible explanations for such puzzling events. Sheep rancher Mary Thoman's difficulties fit this narrative: "So we're just caught here; we can't do anything to defend ourselves. We have to sit here and helplessly watch it [a grizzly] kill our animals. And wait for them [game wardens] to get their act together enough to get it out of here, and by then some more animals suffer. And you gotta go back and see the walking dead; I mean, that's even worse. There's a sheep down in there with her udder eaten off, and here's the little lamb following her, and there's nothing we can do—nothing—just sit here and wait. And then hope that we can find them all to submit the [compensation] claim. That's the hardest part for us. It seems something's gone wrong in this country."[91]

Thoman's comments touch on several important themes: being deprived of property, being prevented from defending one's livestock, bureaucratic entanglements, logistical difficulties of getting compensation, lack of control over any of these events, and a generalized disappointment that such things could happen in America.

The symbolic connection between grizzlies and the idea of a conspiracy to "get rid of us" is another indication of the high level of sociopolitical conflict in the West. Politicians can exploit this conflict by encouraging this kind of worldview and then stepping to the fore to defend everyone against the menace. There are numerous examples of politicians—at the federal, state, and county levels—exerting pressure in overt and covert ways to lessen grizzly recovery efforts. Senator Craig Thomas (R-WY) has a steady record of advocating weaker protection for grizzlies. He has argued for delisting for several years, stating that "by any reasonable measure, the grizzly bears in and around Yellowstone have reached the goals established under the recovery plan and the delisting process should begin."[92] Senator Mike Enzi (R-WY) meets with USFS staff to convey his wishes, including more timber harvest and no expansion of food storage policies.[93] At the state level, a joint committee of the Wyoming legislature sponsored a bill in 1998 to end compensation to ranchers who lost livestock to grizzlies, with the apparent intent of making the situation bad enough to bring about delisting.[94] In 2000 the Wyoming Game and Fish Commission planned to cut in half WGFD's budget for grizzly recovery and management efforts.[95]

On the other side in the polarized sociopolitical context of grizzly bear management are the conservationists. Although many people in western Wyoming apparently believe grizzly bear recovery has gone entirely too far, conservationists believe it has only begun. Many organized conservation interests are not satisfied with the current state of grizzly bear recovery. Their concerns are at least partly fueled by scientific disputes over many aspects of grizzly conservation. For example, methods used to estimate the size of the grizzly population have caused great controversy, leading to an interesting exchange of articles in the academic journals.[96] Prominent grizzly researchers such as John Craighead have weighed in against current management policy in journal articles, opinion pieces, and books.[97]

These concerns led to a lawsuit filed in early 1994 by regional and national conservation organizations against the U.S. Department of the Interior (which oversees the USFWS) over the 1993 Grizzly Bear Recovery Plan.[98] Conservationists challenged the plan on several points, including population goals, protection of habitat, and provisions for reconnecting Yellowstone grizzlies to other grizzly populations. Population goals stated in the 1993 plan include numbers of females with cubs-of-the-year (with a goal of at least 15 over a running 6-year average); the distribution of females with cubs-of-the-year throughout the Recovery Zone; and limiting human-caused mortality of grizzlies to no more than 4 percent of the estimated total population.[99] Fifteen females with cubs-of-the-year correspond to an estimated minimum of 158 bears total in the Greater Yellowstone Ecosystem (in

2001, for comparison, the Interagency Grizzly Bear Study Team identified 42 unique female grizzlies with cubs-of-the year; the team counted 50 unique females with cubs in 2002).[100]

In 1995 a federal judge handed down a mixed decision on the Recovery Plan lawsuit. The decision left USFWS's population goals unchanged, because the Endangered Species Act does not mandate any particular set of quantitative objectives. The decision also found that the act does not require that populations be connected to one another. It did rule against USFWS on habitat protection, finding that the Recovery Plan lacked specificity on how it would conserve grizzly habitat. The decision stated that "the promise of habitat based recovery criteria some time in the future is simply not good enough."[101]

The dispute over population goals, although ostensibly settled in the lawsuit, continues to trouble grizzly recovery today. Many conservationists argue that the current recovery goals are too low and that Yellowstone grizzlies should be reconnected with other grizzly populations farther north. They base these arguments on the long-term genetic and demographic health of the grizzly population.[102] However, "long-term" is a subjective determination and is not defined in the Endangered Species Act. Without an authoritative mandate from Congress, the USFWS has no clear basis to select a particular time horizon for conservation planning. Thus, we may end up with population goals that correspond to a much shorter planning horizon.

Habitat concerns also are a durable source of controversy. Leaving aside disputes over how large a grizzly population we should aim for, conservationists point out that current threats to key grizzly foods in Greater Yellowstone make it unlikely that we can sustain grizzlies at their present level. For example, whitebark pine faces several threats and may be largely eliminated in coming decades. An imported fungus, white pine blister rust, is attacking and killing whitebark pines across the region. Only a small percentage of the trees are resistant to the disease. Compounding this threat, fire exclusion has allowed competing tree species such as subalpine fir to move into whitebark stands and occupy all available space, preventing those few trees surviving blister rust from reproducing. Finally, recent drought conditions have led to a massive outbreak of mountain pine beetles, which bore into the trees' bark to rear young, killing the tree in the process.[103]

As discussed above and as outlined thoroughly in the research of David Mattson, the grizzly population is extraordinarily dependent on whitebark pine.[104] How to treat this threat to the population is a matter of considerable disagreement between conservationists and those who think grizzlies are recovered and should be delisted. Grizzlies are intelligent and adaptive, and individual bears will be able to adjust their feeding habits if whitebark seeds are no longer available. As other researchers point out, some of the ecosystem's grizzlies never eat whitebark seeds anyway.[105] Taken at the population level, however, there is great cause for concern. Overall, many grizzlies may "adapt" to the loss of whitebark pine by foraging at lower elevations, where they are likely to encounter people. Some may "adapt" by moving into human-dominated areas and eating unnatural foods, which would lead to their deaths.

An alternative course is to begin now, before whitebark is devastated, to give bears access to more habitat where they can roam and have minimal conflicts with people. For this reason, many conservationists believe that it is necessary to allow grizzlies to expand into historic range wherever possible. For the most part, this appears to be the rationale for advocating protection of grizzlies. This motive is in sharp contrast to Lois Herbst's claim that grizzly conservation is part of an effort to rid the landscape of rural people. Communication between these two poles appears to be lacking, although there have been some noteworthy efforts. Compounding the problem, there are in fact conservationists whose express mission is to eliminate western public lands grazing. These advocates point to conflicts with grizzlies and other predators as part of the reason for their position.[106]

In summary, grizzly conservation is highly politicized in that various interests see the real, on-the-ground conflicts as part of a much larger political fight. In such a situation, wholesale sweeping improvements seem unlikely. Decision makers who could initiate such improvements may either fear the controversy or actively exploit it for their own ends.

Management Options: A Participatory Strategy for Improving Grizzly Conservation

Grizzly conservation is a formidable technical challenge, considering the vast area, rough terrain, and secretive nature of the bears themselves. The challenge seems to be utterly overwhelming when we take into account the complexity of the larger sociopolitical context. History shows that our ability to devise large-scale, comprehensive solutions to messy, value-laden, and symbolically charged problems like grizzly conservation is severely limited. We simply are not cognitively capable of dealing with this amount of complexity. Moreover, even if an appropriate broad-scale strategy could be designed, decision makers would be unlikely to bring it into effect. Rather than ignoring these limitations and constraints with ill-fated attempts at comprehensive problem solving, we should instead frame problems in a way and at a scale that we can grapple with them practically.[107] We recommend a strategy of site-specific problem-solving initiatives that use "dialogue," or reasoning together among local citizens, to deal with the challenges of grizzly conservation and create opportunities for progress. The objectives of this strategy are to alleviate immediate grizzly-human conflicts while building social capacity and disconnecting the symbolic ties between grizzly conservation and broader sociopolitical issues.

Concentrating efforts at the local level takes advantage of on-the-ground opportunities to solve problems. It also offers a way to manage the cultural and political complexity of grizzly conservation by dealing with a manageably sized portion of participants in the policy process. Substantive issues such as public safety and land use regulations can be addressed through participatory processes that, in themselves, mitigate sociopolitical conflict by giving people an active role in problem solving.[108]

These small-scale problem-solving efforts should not be atomistic or disjointed. A network of site-specific grizzly conservation programs would facilitate development and testing of innovative ideas in different locations. With built-in flexibility and multivariate monitoring, feedback from multiple, concurrent, localized conservation projects would emerge rapidly, allowing replication of successes and course corrections when necessary. Problem-solving innovations from one area could then be applied to other areas. The alternatives to such an approach are unpromising attempts at large-scale, comprehensive planning, or disjointed responses to localized problems, making no use of learning networks or experience in similar cases.[109]

As we outlined above, grizzly conservation carries a heavy symbolic burden. Many citizens apparently view the issue as an "us versus them" struggle, pitting the people of the West against a hostile, alien culture. Senator Thomas captured this view: "You hear quite a different tune from the people who live in the area than from the people who live in Maryland and think grizzly bears ought to be everywhere."[110] If people view grizzlies as foot soldiers sent by a distant enemy, it seems unlikely that these people will welcome efforts to keep grizzlies around in quiet coexistence. In the following sections we describe what we mean by dialogue, why this approach is well suited to grizzly conservation, and how a strategy built on dialogue could be structured.

Dialogue as a Specific Practice

We recommend dialogue as both a first step and as a mechanism for continuous guidance. Daniel Yankelovich provides a succinct and accessible overview of dialogue as a particular procedure, and not as a colloquial synonym for collaboration, negotiation, or consensus. Dialogue, in Yankelovich's account, is a process of reasoning together. Its features include equality among participants, empathetic listening and respect for others, and an obligation to make one's reasoning and assumptions as clear as possible. If we explain and explore the assumptions behind our opinions, we may discover that we have clung to beliefs and positions that have little factual basis. Empathetic listening and treating each other as equals may allow reason-based collective opinions to be formed. Yankelovich stated, "In genuine dialogue, there is no arm-twisting, no pulling of rank, no hint of sanctions for holding politically incorrect attitudes, no coercive influences of any sort, whether overt or indirect."[111]

Although the concept of dialogue may seem idealistic, there are compelling reasons to employ dialogue to solve complex sociopolitical problems. Beliefs and values cannot be readily and reliably measured and used to craft effective technical solutions. Even if these variables could be used in this way, technical-rational solutions imposed from above only perpetuate a lack of involvement and sense of powerlessness among citizens, which can block implementation. For example, in a review of agency land management planning processes, the USFS Policy Analysis staff concluded that attempts to "resolve social and political problems solely with technical solutions . . . [alienated] many whose continued participation we needed in the future."[112]

The solution to these difficulties is to engage citizens actively so that they have the opportunity to represent their own opinions accurately and integrate their values with those of others. Working collaboratively with relevant specialists and skilled facilitators, citizens could be involved directly in designing technical solutions while resolving value conflicts among themselves. This process seeks reason through intersubjective communication, as opposed to the subject-object reason pursued by technical-rational problem solvers. That is, reason regarding social problems is a collective endeavor. Processes of this sort seek to form reason-based public opinions, as opposed to merely measuring baseline opinions in survey projects.[113]

Citizen-driven processes for developing reason-based opinion have many advantages. They offer the prospect of developing an empowered, engaged, and informed citizenry. They also represent a promising alternative to dysfunctional mechanisms for solving complex problems in which scientific uncertainty is combined with value conflicts. But it is important to acknowledge the challenges and limitations of such approaches and to evaluate their applicability on a case-by-case basis. Thus, it is appropriate to ask whether these processes would actually improve grizzly conservation.

Applicability of Dialogue to Grizzly Conservation

There are several reasons that grizzly conservation is amenable to dialogue processes. First, successful coexistence between people and grizzlies requires many individuals to integrate bear-safe behaviors into their everyday lives. People will have to change their behavior while camping, hiking, hunting, and in many cases around their homes. Unless there is political support and resources for intensive monitoring and stringent enforcement, implementing coexistence practices must rely in good part on voluntary compliance and public goodwill. Voluntary compliance and informal enforcement (e.g., peer pressure) are more likely if citizens are actively involved in developing programs. Second, grizzly conservation is highly polarized at broader scales, as we discussed above. At these scales, people are unlikely to have the time or attention to analyze and solve specific, localized grizzly conflicts. Instead, they are more likely to emphasize the divisive, symbolic dimensions of grizzly conservation.

Third, it is not clear that pursuing improved grizzly conservation through national-scale initiatives makes sense. Political scientist Stephen Meyer described the general problem: "Although the public clearly has signed onto the agenda of protecting biodiversity the issue has—much to the chagrin of environmentalists—very low political salience. It does not mobilize the electorate to political action."[114] Locals, on the other hand, have both the motivation and the knowledge to tackle site-specific management challenges. A small, cohesive group (e.g., ranchers) that stands to suffer clear costs is much more likely to mobilize than diffuse, nominal groups (e.g., suburbanites on either coast) that may be only fleetingly concerned with grizzly conservation.[115] Yet it is possible and practical to structure local dialogue processes to incorporate national interests. Many national environmental groups have staff in the Yellowstone area, and regional groups have offices there too. Paid staff members could participate in dialogues, or local residents who belong to such organizations could play a role. National interests are also represented through the provisions of applicable federal legislation such as the Endangered Species Act.

Additionally, local processes have many practical benefits. Site-specific programs that have diverse funding bases and strong local support would be resilient in the face of systemic changes in politics and budgets, such as those we are experiencing now. These programs are also far less expensive to maintain and monitor than "top down," regulatory approaches. Successful projects would demonstrate that cooperation is possible, changing the behavior of reluctant allies and strident opponents. Demonstrated success could in turn attract more funding and other resources, making the next projects easier.[116]

Practical Recommendations for Making Progress

Dialogue-driven processes for resolving grizzly conservation problems must develop on a case-by-case basis. Policy researchers Ron Brunner and Christy Colburn pointed out that such efforts "proceed within the distinctive opportunities and constraints of a place-based community."[117] One promising structure would be for a relatively uncontroversial nongovernmental organization to convene meetings on a trial basis in areas that experience grizzly-human conflicts. The organization would need to have the appropriate financial and political resources to initiate such a process. These resources would include meeting space, travel costs, information resources, and skilled facilitators.[118]

As a first step, efforts should concentrate on readily identifiable problem areas.[119] Annual reports on grizzly-human conflicts in the Greater Yellowstone Ecosystem provide clear information about where problem areas are and should provide some ability to forecast future conflicts.[120] Although forecasting movements of individual grizzlies is unrealistic, the human contribution to these conflicts is similar across cases. With predictive models of where bears are likely to expand to next (based on current distribution of bears, terrain features, and available natural foods) and information about the distribution and habits of people, it is feasible to identify areas where new conflicts are likely to erupt. Problem areas tend to be relatively localized, involving distinct human communities.

Localized problem-solving groups in such communities would identify their own priorities and focal points. Some possible roles would include designing and implementing effective, cross-jurisdictional food storage and sanitation programs. Keeping bears out of garbage and anthropogenic foods is a straightforward way of minimizing conflicts. Yet, such programs must be thorough—it takes very few experiences with human foods or garbage for a wild bear to become a problem bear. Coordinating systemic efforts across federal, state, local, and private jurisdictions is a major challenge. Local citizens, having a network of social relationships and detailed knowledge of the area, may be uniquely well suited to this task.

Ideally, all affected or interested local citizens would become involved. However, there is only a limited pool of citizens who have the time, inclination, and disposition to participate constructively. Many of these people are already involved in all the civic activities they can handle. Many agricultural people, whose involvement would be crucial, have little discretionary time. Expecting citizens themselves to be the sole "keepers" of a process is unrealistic, thus the need for a nongovernmental organization or other entity to perform this role.

Local groups might be able to find points of entry into broader scale policy making as well. For example, WGFD has proposed developing a public-private endowment to provide permanent funding for large predator conservation.[121] This endowment remains in a conceptual stage, but may well be worth pursuing. Localized groups could identify funding needs, help persuade decision makers to allocate public funds to the endowment, and also assist with raising private funds. An endowment funded by an allocation from the U.S. Treasury would ensure an operational minimum of funding for grizzly conservation every year, reducing programmatic vulnerabilities to politically motivated budget cuts and maintaining consistent efforts. Moreover, such a fund would help spread the burdens of predator conservation to the nation at large, instead of imposing them primarily on residents of the rural West.[122]

A second focus of broader policy-making activity would be to improve livestock compensation efforts. As we described earlier, existing compensation programs are better than nothing, but could work far better than they do. The logistical and administrative hurdles of documenting predator-killed livestock and then getting compensation can be daunting. Many kills go undetected because of rough terrain and consumption of all evidence. Recognizing this, the Wyoming Game and Fish Commission revised its grizzly predation compensation program in 2003 to reimburse for missing livestock, as long as a rancher has suffered some documented losses to grizzlies.[123] Further, livestock operations are not reimbursed for the time they spend in documenting kills, nor are they compensated for the stress and weight loss that predators may impose on livestock that are not directly attacked. Some have proposed insurance pools that would pay for losses without direct documentation of each predator kill.[124] Citizens' problem-solving groups could try out these new ideas on a limited basis.

These efforts, spread out over large areas and involving many citizens, may yield many effective solutions and problem-solving models. Although diverse efforts on small scales offer many advantages, it will be important to ensure communication and coordination among them. On the scale of individual projects, coordination with others is vital for several reasons. First, people undertaking a project should avail themselves of the trial-and-error lessons from similar projects to avoid needlessly repeating mistakes or wasting resources. Examples of this abound in grizzly conservation, whether it is finding reliable suppliers of bear-resistant garbage cans or properly constructing an electric fence around sheep pasture. Second, grizzlies often range over considerable distances. A community may take extensive precautions to keep bears out of anthropogenic foods, yet neighboring communities in the next watershed may fail to take such measures. The range of local grizzly bears may span both communities. The result could be that the community that took the effort to avoid problems may still be plagued by dangerous bears that learned bad habits elsewhere.[125]

On a broader scale, coordination is a "force multiplier" for improving practices throughout grizzly range. On a functional level, coordination can ensure uniformly effective programs over appropriately large scales for successful grizzly conservation. Coordinated efforts can also help to identify those areas in need of additional attention and funds and then help to supply those resources. On the level of symbolic politics, the ability to publicize successful projects will be key in helping weaken the connection between negative symbols and the actual grizzlies that live in the region. Successful models would demonstrate that traditional Westerners can cooperate with proponents of grizzly conservation, rather than being locked in an "us-versus-them" struggle of identity politics. Reluctant allies and strident opponents may then change their behavior. Demonstrated successes could in turn attract more funding and other resources, making the next projects easier to implement.[126] Coordination across projects, then, is a vital task. It will be necessary for some organization to take responsibility for this function, rather than just assuming that it will happen spontaneously. Web sites, annual symposia, active outreach, and other methods of communication will ensure that communication, coordination, and learning will take place.

Conclusion

Grizzlies have made a remarkable comeback in Greater Yellowstone. The challenge now will be managing this success for long-term sustainability. Grizzlies in this region are important ecologically and culturally. These bears are an integral part of the Yellowstone experience for many Americans. For many years, we apparently valued the park's grizzlies as a source of entertainment, evidenced by the bear feeding shows that continued up to the 1940s.[127] As our ecological knowledge has expanded and our environmental ethics have matured, our culture has perhaps grown to value wild, self-perpetuating populations of wildlife. With this value comes the realization that to be a wild population Yellowstone's grizzlies must be regarded as Greater Yellowstone Ecosystem grizzlies. They must have access to an area much larger than the park's 2.2 million acres if they are to survive and thrive through the coming centuries. The present dire condition of whitebark pine underscores this imperative.

As we have seen, however, giving grizzlies access to a larger portion of the region is no simple matter. Perhaps the most important aspect to recognize in the current conflicts is the degree to which grizzly conservation is burdened by symbolic linkages to larger, sometimes unrelated issues. It is clear that there are real, practical problems that come with recolonizing grizzlies and that people are right to be alarmed at these problems. Unfortunately, the metamorphosis of *U. arctos* into a potent political symbol has dramatically impaired our collective ability to talk about, reason through, and deal with the practical problems of grizzly-human coexistence.

In this chapter, then, we have outlined the challenges that we see in Greater Yellowstone grizzly conservation. Having a clear understanding of the problem to be solved is essential if we are to choose the proper tools and techniques. We believe that problem-solving structures should be put in place to deal with both the practical problems and with the symbolic problems. Small-scale, yet well-coordinated processes for collective reasoning and practical progress are the foundation of our approach. There are many details remaining to be worked out—for example, who will take the lead in initiating and coordinating these processes—but we believe this direction is well worth pursuing.

References

1. T. W. Clark and D. Casey, 1992, *Tales of the grizzly,* Homestead Publishing, Moose, WY.

2. See http://geonames.usgs.gov for a searchable database of place names.

3. F. C. Craighead, 1979, *Track of the grizzly,* Sierra Club, San Francisco.

4. K. A. Gunther, M. T. Bruscino, S. Cain, L. Hanauska-Brown, K. Frey, M. A. Haroldson, and C. C. Schwartz, 2002, "Grizzly bear human conflicts and management actions in the greater Yellowstone ecosystem 2001," 56–93 in C. C. Schwartz and M. A. Haroldson, eds., *Yellowstone grizzly bear investigations: Annual report of the Interagency Grizzly Bear Study Team, 2001,* U.S. Geological Survey, Bozeman, MT.

5. See H. D. Lasswell and M. S. McDougal, 1992, *Jurisprudence for a free society,* v. II, Kluwer, Dordrecht, The Netherlands, 1052.

6. R. Huntington, 1999, "Bear-ly making ends meet," *Jackson Hole Guide,* September 15, A1, 23; S. Pyare, S. Cain, D. Moody, C. Schwartz, and J. Berger, 2004, "Carnivore re-colonisation: Reality, Possibility, and a non-equilibrium century for grizzly bears in the Southern Yellowstone Ecosystem," *Animal Conservation* 7, 1–7.

7. "Grizzly" is a regional name typically applied to interior North American *U. arctos;* coastal members of the species are often called "brown bears."

8. D. J. Mattson, 2000, "Brown bear," 66–71 in R. P. Reading and B. Miller, eds., *Endangered animals: Reference guide to conflicting issues,* Greenwood, Westport, CT.

9. I. Chestin, 1999, "Status and management of the brown bear in Russia," 136–143 in C. Servheen, S. Herrero, and B. Peyton, comp., *Bears: Status survey and conservation action plan,* IUCN, Gland, Switzerland.

10. Servheen et al., *Bears.* There are far more American black bears than there are numbers of the other seven bear species combined.

11. L. P. Waits, S. L. Talbot, R. H. Ward, and G. F. Shields, 1998, "Mitochondrial DNA phylogeography of the North American brown bear and implications for conservation," *Conservation Biology* 12, 408–417.

12. Waits et al., "Mitochondrial DNA phylogeography of the North American brown bear and implications for conservation."

13. F. L. Craighead, M. E. Gilpin, and E. R. Vyse, 1999, "Genetic considerations for carnivore conservation in the greater Yellowstone ecosystem," ch. 11 in T. W. Clark et al., eds., *Carnivores in ecosystems: The Yellowstone experience,* Yale, New Haven. See also M. E. Soulé, 1980, "Thresholds for survival: Maintaining fitness and evolutionary potential," ch. 9 in B. A. Wilcox and M. E. Soulé , eds., *Conservation biology: An evolutionary-ecological perspective,* Sinauer, Sunderland, MA.

14. Waits et al., "Mitochondrial DNA phylogeography of the North American brown bear and implications for conservation," 416.

15. Mattson, "Brown bear"; J. J. Craighead, J. S. Sumner, and J. A. Mitchell, 1995, *The grizzly bears of Yellowstone: Their ecology in the Yellowstone ecosystem, 1959–1992,* Island Press, Washington, D.C.; K. D. Rode, C. T. Robbins, and L. A. Shipley, 2001, "Constraints on herbivory by grizzly bears," *Oecologia* 128, 62–71.

16. D. J. Mattson, 1990, "Human impacts on bear habitat use," *International Conference on Bear Research and Management* 8, 33–56

17. Craighead et al., *The grizzly bears of Yellowstone;* C. Servheen, 1989, "The management of the grizzly bear on private lands: Some problems and possible solutions," 195–200 in M. Bromley, ed., *Bear-people conflicts: Proceedings of a symposium on management strategies,* Northwest Territories Department of Renewable Resources, Yellowknife; Mattson, "Human impacts on bear habitat use"; K. A. Gunther, 1994, "Bear management in Yellowstone National Park, 1960–93," *International Conference on Bear Research and Management* 9, 549–560.

18. Craighead et al., *The grizzly bears of Yellowstone;* J. L. Weaver, P. C. Paquet, and L. F. Ruggiero, 1996, "Resilience and conservation of large carnivores in the Rocky Mountains," *Conservation Biology* 10, 964–976.

19. D. J. Mattson, R. R. Knight, and B. M. Blanchard, 1986, "The effects of development and primary roads on grizzly bear habitat use in Yellowstone National Park, Wyoming," in P. Zager, ed., *Bears: Their biology and management,* International Association for Bear Research and Management, Washington, D.C.

20. B. N. McLellan and D. M. Shackelton, 1988, "Grizzly bears and resource-extraction industries: Effects of roads on behaviour, habitat use and demography," *Journal of Applied Ecology* 25, 458.

21. S. Herrero, 1985, *Bear attacks: Their causes and avoidance,* Lyons and Buford, New York.

22. D. J. Mattson and M. Reid, 1991, "Conservation of the Yellowstone grizzly bear," *Conservation Biology* 5, 364–372.

23. Mattson et al., "The effects of development and primary roads on grizzly bear habitat use in Yellowstone National Park, Wyoming."

24. K. A. Gunther, M. T. Bruscino, S. Cain, L. Hanauska-Brown, K. Frey, M. A. Haroldson, and C. C. Schwartz, 2002, "Grizzly bear-human conflicts and management actions in the Greater Yellowstone Ecosystem 2001," 56–93 in C. C. Schwartz and M. A. Haroldson, eds., *Yellowstone grizzly bear investigations: Annual report of the Interagency Grizzly Bear Study Team, 2001,* U.S. Geological Survey, Bozeman, MT.

25. USFWS, 1993, *Grizzly bear recovery plan,* USFWS, Missoula, MT.

26. Clark and Casey, *Tales of the grizzly.*

27. D. J. Mattson and T. Merrill, "Extirpations of grizzly bears in the contiguous United States, 1850–2000," *Conservation Biology* 16, 1123–1136.

28. USFWS, *Grizzly bear recovery plan.*

29. Craighead et al., *The grizzly bears of Yellowstone.*

30. A. K. Wondrak, 2002, "Wrestling with Horace Albright, Part I: Edmund Rogers, visitors, and bears in Yellowstone National Park," *Magazine of Western History* 52(3), 2–15.

31. P. Schullery, 1992, The bears of Yellowstone, High Plains, Boulder, CO.

32. A. Murie, 1948, "Cattle on grizzly bear range," *Journal of Wildlife Management* 12, 57–72. The Teton National Forest was combined with the Bridger National Forest in 1973 to form the Bridger-Teton National Forest.

33. Pyare et al., "Carnivore re-colonisation: Reality, possibility, and a non-equilibrium century for grizzly bears in the Southern Yellowstone Ecosystem."

34. D. L. Condin, 1956, "The Yellowstone grizzly," *Wyoming Wildlife,* Dec., 8–15, quote on 11.

35. G. W. Reynolds, 1959, "The silver ghost," *Wyoming Wildlife,* July, 21–23.

36. D. S. Moody, D. Hammer, M. Bruscino, D. Bjornlie, R. Grogan, and B. Debolt, 2002, *Wyoming grizzly bear management plan,* Wyoming Game and Fish Department, Cheyenne.

37. Clark and Casey, *Tales of the grizzly;* Craighead, *Track of the grizzly;* Schullery, *The bears of Yellowstone.*

38. Craighead et al., *The grizzly bears of Yellowstone;* Pyare et al., "Carnivore re-colonisation: Reality, possibility, and a non-equilibrium century for grizzly bears in the Southern Yellowstone Ecosystem."

39. Craighead, *Track of the grizzly.*

40. Craighead, *Track of the grizzly;* Schullery, *The bears of Yellowstone.*

41. Gunther, "Bear management in Yellowstone National Park, 1960–93."

42. Craighead et al., *The grizzly bears of Yellowstone.*

43. Craighead, *Track of the grizzly.*

44. J. J. Craighead, K. R. Greer, R. R. Knight, and H. I. Pac, 1988, *Grizzly bear mortalities in the Yellowstone ecosystem, 1959–1987,* Montana Department of Fish, Wildlife and Parks, Bozeman, MT; Craighead, *Track of the grizzly.*

45. Moody et al., *Wyoming grizzly bear management plan.*

46. M. Meagher and J. R. Phillips, 1983, "Restoration of natural populations of grizzly and black bears in Yellowstone National Park," *International Conference on Bear Research and Management* 5, 152–158; Craighead et al., *Grizzly bear mortalities in the Yellowstone Ecosystem, 1959–1987.*

47. http://nrmsc.usgs.gov/research/igbst-home.htm.

48. National Academy of Sciences, Report of the Committee on Yellowstone Grizzlies, July 1974, quote on 38; see also F. Craighead, *Track of the grizzly.*

49. *Endangered Species Act,* 16 USC 1532 [3].

50. S. L. Yaffee, 1982, *Prohibitive policy: Implementing the federal Endangered Species Act,* MIT Press, Boston.

51. D. J. Mattson and J. J. Craighead, 1994, "The Yellowstone grizzly bear recovery program: Uncertain information, uncertain policy," 101–130 in T. W. Clark, R. P. Reading, and A. L. Clarke, editors, *Endangered species recovery: Finding the lessons, improving the process,* Island Press, Washington, D.C.

52. Craighead et al., *The grizzly bears of Yellowstone,* see population estimates on 297.

53. Meagher and Phillips, "Restoration of natural populations of grizzly and black bears in Yellowstone National Park."

54. Craighead et al., *The grizzly bears of Yellowstone.*

55. R. Wauer, 1982, Memorandum to Interagency Grizzly Bear Steering Committee, August 17.

56. *New York Times,* 1982, "Grizzlies seen as imperiled in Wyoming," Oct. 10, A80.

57. USFWS, *Grizzly bear recovery plan.*

58. M. D. Strickland, 1990, "Grizzly bear recovery in the contiguous United States," *International Conference on Bear Research and Management* 8, 5–9.

59. Strickland, "Grizzly bear recovery in the contiguous United States."

60. D. J. Mattson, B. M. Blanchard, and R. R. Knight, 1991, "Food habits of Yellowstone grizzly bears, 1977–1987," *Canadian Journal of Zoology* 69, 1619–1629.

61. Gunther, "Bear management in Yellowstone National Park, 1960–93."

62. C. M. Pease and D. J. Mattson, 1999, "Demography of the Yellowstone grizzly bears," *Ecology* 80, 957–975.

63. D. J. Mattson, 2000, *Causes and consequences of dietary differences among Yellowstone grizzly bears* (Ursus arctos), Ph.D. dissertation, University of Idaho, Moscow.

64. D. J. Mattson, B. M. Blanchard, and R. R. Knight, 1992, "Yellowstone grizzly bear mortality, human habituation, and whitebark pine seed crops," *Journal of Wildlife Management* 56, 432–442.

65. Pease and Mattson, "Demography of the Yellowstone grizzly bears;" C. C. Schwartz, M. A. Haroldson, K. A. Gunther, and D. Moody, 2002, "Distribution of grizzly bears in the greater Yellowstone ecosystem, 1990–2000," *Ursus* 13, 203–212.

66. Mattson, *Causes and consequences of dietary differences among Yellowstone grizzly bears* (Ursus arctos).

67. B. Loomis, 1992, "Grizzly bear plagues Walton cattle," *Jackson Hole Guide,* October 14, A1, A13.

68. See R. R. Knight and L. L. Eberhardt, 1985, "Population dynamics of Yellowstone grizzly bears," *Ecology* 66, 323–334. Knight and Eberhardt classify Blackrock-Spread Creek and adjacent habitat as "high" or "highest density" for grizzlies.

69. A. M. Thuermer, Jr., 1993, "Togwotee griz gets a reprieve," *Jackson Hole News,* October 13, A1, 19.

70. For a critique of compensation programs, see C. Urbigkit, 2004, "The compensation fraud," *Sublette Examiner,* 22 January, http://www.subletteexaminer.com/archives/01_22_04_top_stories.shtml (accessed July 15, 2004).

71. Loomis, "Grizzly bear plagues Walton cattle," quote on A13.

72. B. Loomis, 1993, "Two or more grizzlies feed on private cattle," *Jackson Hole Guide,* July 21, A19.

73. Thuermer, "Togwotee griz gets a reprieve," A2.

74. Loomis, "Grizzly bear plagues Walton cattle," A13.

75. Loomis, "Grizzly bear plagues Walton cattle"; see also various letters from conservationists to USFS, on file with senior author.

76. M. Riley, 1993, "Debate heats up over grizzly status," *Casper Star-Tribune,* October 18, B1.

77. Washakie County Commissioners, 1993, letter to President Clinton, Worland, WY, on file with senior author.

78. The sources for these statistics are the annual conflict summaries published by the IGBC. K. A. Gunther, K. Aune, S. Cain, T. Chu, and C. M. Gillin, 1993, *Grizzly bear-human conflicts in the Yellowstone Ecosystem 1992,* IGBC, Yellowstone National Park, WY; K. A. Gunther, M. Bruscino, S. Cain, T. Chu, K. Frey, and R. R. Knight, 1994, *Grizzly bear-human conflicts, confrontations, and management actions in the Yellowstone Ecosystem 1993,* IGBC, Yellowstone National Park, WY; K. A. Gunther, M. Bruscino, S. Cain, T. Chu, K. Frey, and R. R. Knight, 1995, *Grizzly bear-human conflicts, confrontations, and management actions in the Yellowstone Ecosystem 1994,* IGBC, Yellowstone National Park, WY; K. A. Gunther, M. Bruscino, S. Cain, T. Chu, K. Frey, and R. R. Knight, 1996, *Grizzly bear-human conflicts, confrontations, and management actions in the Yellowstone Ecosystem 1995,* IGBC, Yellowstone National Park, WY; K. A. Gunther, M. Bruscino, S. Cain, T. Chu, K. Frey, and R. R. Knight, 1997, *Grizzly bear-human conflicts, confrontations, and management actions in the Yellowstone Ecosystem 1996,* IGBC, Yellowstone National Park, WY; K. A. Gunther, M. Bruscino, S. Cain, T. Chu, K. Frey, M. A. Haroldson, and C. C. Schwartz, 1998, *Grizzly bear-human conflicts, confrontations, and management actions in the Yellowstone Ecosystem 1997,* IGBC, Yellowstone National Park, WY; K. A. Gunther, M. Bruscino, S. Cain, J. Copeland, K. Frey, M. A. Haroldson, and C. C. Schwartz, 1999, *Grizzly bear-human conflicts, confrontations, and management actions in the Yellowstone Ecosystem 1998,* IGBC, Yellowstone National Park,

WY; K. A. Gunther, M. Bruscino, S. Cain, J. Copeland, K. Frey, M. A. Haroldson, and C. C. Schwartz, 2000, *Grizzly bear-human conflicts, confrontations, and management actions in the Yellowstone Ecosystem 1999,* IGBC, Yellowstone National Park WY; K. A. Gunther, M. Bruscino, S. Cain, J. Copeland, K. Frey, M. A. Haroldson, and C. C. Schwartz, 2001, *Grizzly bear-human conflicts, confrontations, and management actions in the Yellowstone Ecosystem 2000,* IGBC, Yellowstone National Park, WY; K. A. Gunther, M. Bruscino, S. Cain, L. Hanauska-Brown, K. Frey, M. A. Haroldson, and C. C. Schwartz, 2002, *Grizzly bear-human conflicts, confrontations, and management actions in the Yellowstone Ecosystem 2001,* IGBC, Yellowstone National Park WY.

79. C. M. Cromley, 2000, "The killing of grizzly bear 209: Identifying norms for grizzly bear management," 173–220 in T. W. Clark et al., eds., *Foundations of natural resources policy and management,* Yale University Press, New Haven.

80. B. Israelsen, 2003, "Room for bears," *Salt Lake Tribune,* 4 September, http://www.sltrib.com/search (accessed 07 September 2003).

81. Annual IGBC conflict summaries 1993–2001; C. C. Schwartz, M. A. Haroldson, K. A. Gunther, and D. Moody, 2002, "Distribution of grizzly bears in the greater Yellowstone ecosystem, 1990–2000," *Ursus* 13, 203–212.

82. Huntington, "Bear-ly making ends meet."

83. R. R. Knight and S. Judd, "Grizzly bears that kill livestock," *International Conference on Bear Research and Management 5,* 186–190; also M. Bruscino and B. DeBolt, 2001, letter to Wyoming Game and Fish Department Director John Baughman, January 6.

84. C. Urbigkit, 1999, "Stock killer destroyed," *Pinedale Roundup,* July 15, 1, 28; R. Odell, 1999, "Groups seek changes in griz killing policy," *Jackson Hole News,* August 7, A38–A39; R. Huntington, 1999, "Predator troubles down for Upper Green sheep," *Jackson Hole Guide,* October 13, A1.

85. Urbigkit, "The compensation fraud."

86. Cromley, "The killing of grizzly bear 209."

87. See http://www.nps.gov/yell/nature/animals/bear/infopaper/info1.html.

88. J. Haines, 2002, "Wyoming county bans grizzly bears," *Bozeman Daily Chronicle,* May 4, A3.

89. J. Haines, "Wyoming county bans grizzly bears," A3.

90. R. Menzies, "Bear tracks and bear facts," *Range,* http://www.rangemagazine.com/archive/stories/winter01/bear_tracks.htm.

91. R. Menzies, 1999, "Grizzly picnic," in *Range,* Winter, 4–8, quote on 5.

92. Quoted in T. Wilkinson, 1998, "Grizzly war," *High Country News,* Nov. 9.

93. Senator Mike Enzi, news release dated May 1, 2002, http://enzi.senate.gov/grizzl.htm.

94. Associated Press, 1998, "Panel seeks to end grizzly bear damage payments," *Casper Star-Tribune,* October 9, B1.

95. Associated Press, 2000, "Game and Fish tentatively cuts grizzly funding in half," *Casper Star-Tribune,* April 29.

96. D. J. Mattson, 1997, "Sustainable counts of grizzly bear mortality calculated

from counts of females with cubs-of-the-year: An evaluation," *Biological Conservation* 81, 103–111.

97. J. J. Craighead et al., *The grizzly bears of Yellowstone;* J. J. Craighead, 1995, "Grizzly delisting push lacks data," *Casper Star-Tribune,* January 19, B1; J. J. Craighead, 1998, "Status of the Yellowstone grizzly population: Has it recovered, should it be delisted," *Ursus* 10, 597–602.

98. D. Whipple, 1994, "3 groups to sue over grizzly plan," *Casper Star-Tribune,* January 27, B1.

99. USFWS, *Grizzly bear recovery plan.*

100. C. C. Schwartz and M. A. Haroldson, eds., 2002, *Yellowstone grizzly bear investigations: Annual report of the Interagency Grizzly Bear Study Team, 2001,* U.S. Geological Survey, Bozeman, MT; C. C. Schwartz and M. A. Haroldson, eds., 2002, *Yellowstone grizzly bear investigations: Annual report of the Interagency Grizzly Bear Study Team, 2002,* U.S. Geological Survey, Bozeman, MT.

101. *Audubon v. Babbitt,* U.S. District Court for the District of Columbia, September 29, 1995.

102. See L. L. Willcox and D. Ellenberger, 2000, *The bear essentials for recovery: An alternative strategy for long-term restoration of Yellowstone's great bear,* Sierra Club, Bozeman, MT, for a good summary of conservationists' demands.

103. Whitebark pine research is a growing field. See various chapters in D. F. Tomback, S. F. Arno, and R. F. Keane, eds., 2001, *Whitebark pine communities: Ecology and restoration,* Island Press, Washington, D.C.

104. Mattson, *Causes and consequences of dietary differences among Yellowstone grizzly bears.*

105. Associated Press, 2003, "Experts disagree on bear hunger," *Helena Independent Record,* May 25, www.helenair.com/articles/2003/05/25/montana/a01052503_04.prt.

106. See G. Weurthner and M. Matteson, eds., 2002, *Welfare ranching: The subsidized destruction of the American West,* Island Press, Washington, D.C., for the definitive statement.

107. C. E. Lindblom and E. J. Woodhouse, 1993, *The policy-making process,* 3rd ed., Prentice-Hall, Englewood Cliffs, NJ.

108. G. Borrini-Feyerabend, 1997, *Beyond fences: Seeking social sustainability in conservation,* IUCN, Gland, Switzerland, http://www.iucn.org/themes/spg/Files/beyond_fences/beyond_fences.html.

109. R. D. Brunner and C. H. Colburn, 2002, "Harvesting experience," 201–247 in R. D. Brunner, C. H. Colburn, C. M. Cromley, R. A. Klein, and E. A. Olson, eds., *Finding common ground: Governance and natural resources in the American West,* Yale University Press, New Haven.

110. National Public Radio, 1999, "Weekend Edition Saturday," August 21 (transcript on file with the author).

111. D. Yankelovich, 1999, *The magic of dialogue: Transforming conflict into cooperation,* Simon and Schuster, New York, 42.

112. USFS, Policy Analysis staff, 1990, *Synthesis of the critique of land management planning,* USFS, Washington, 9.

113. Y. Haila, "Environmental problems, ecological scales and social deliberation"; Dryzek, *Deliberative democracy and beyond.*

114. S. M. Meyer, 2001, "Community politics and endangered species protection," 138–165 in J. F. Shogren and J. Tschirhart, eds., *Protecting endangered species in the United States,* Cambridge Press, London, quote on 142.

115. J. T. Heinen and B. S. Low, 1992, "Human behavioural ecology and environmental conservation," *Environmental Conservation* 19, 105–116.

116. D. Jehl, 2003, "On rules for environment, Bush sees a balance, critics a threat," *New York Times,* February 23 (www.nytimes.com/2003/02/23/science/23ENVI.html); Borrini-Feyerabend, *Beyond fences.*

117. Brunner and Colburn, "Harvesting experience," 201.

118. Brunner and Colburn, "Harvesting experience."

119. S. A. Primm, 1996, "A pragmatic approach to grizzly bear conservation," *Conservation Biology* 10, 1026–1035.

120. Gunther et al., 2002, "Grizzly bear human conflicts and management actions in the greater Yellowstone ecosystem."

121. Moody et al., *Wyoming grizzly bear management plan.*

122. T. J. Kaminski, J. F. Gore, and W. D. Owens, 1995, "Trust for wolf conservation: A proposed model for funding carnivore conservation in the US and Canada, 137–141 in A. P. Curlee, A. Gillesberg, and D. Casey, eds., *Greater Yellowstone predators: Ecology and conservation in a changing landscape,* Northern Rockies Conservation Cooperative and Yellowstone National Park, Jackson, WY.

123. J. Gearino, 2003, "Producers to get more for grizzly kills," *Casper Star-Tribune,* August 4; Urbigkit, "The compensation fraud." Under the revised compensation program, ranchers with verified grizzly losses are paid 3.5 times their verified losses to account for calves or sheep lost to unknown causes. This applies only on allotments in rough terrain where verification would be difficult or impossible.

124. See F. H. Wagner, 1972, *Coyotes and sheep: Some thoughts on ecology, economics, and ethics,* Utah State University, Logan, for a discussion of livestock insurance systems.

125. Numerous bear management cases illustrate this problem. See IGBC annual conflict summaries.

126. K. E. Weick, 1984, "Small wins: Redefining the scale of social problems," *American Psychologist* 39, 40–49.

127. Wondrak, "Wrestling with Horace Albright, part I."

Chapter **5**

Wolf Restoration: A Battle in the War over the West

Jason Wilmot and Tim W. Clark

On October 28, 2002, the Wyoming Game and Fish Commission held a public comment forum in Jackson, Wyoming, regarding its draft wolf management plan. Assuming that the state would manage the species after its removal from federal protection under the Endangered Species Act, the plan designated wolves as predators throughout most of the state. This classification would allow unregulated killing of wolves outside the national parks and the wilderness areas adjacent to Yellowstone. In response to this plan, the public voiced a wide range of opinions regarding how the state should manage wolves. Wolf advocate Pete Barry said it is only "a tiny minority of people, mainly hunters and ranchers," who don't want wolves. "Let's not cater to a tiny minority and represent the broad spectrum of America," he said. Outfitter Maury Jones, on the other hand, supported the state's position, saying, "[The federal government has been] cramming things down our throats for too long. We need a Commission and Legislature with a backbone."[1] Comments like these exemplify the contentious nature of wolf recovery and management in the Northern Rockies, and they show the difficulty wildlife managers have in meeting entrenched and opposing demands.

Following the public comment period, Wyoming wildlife officials voted to classify the wolf as a predator in most of the state, despite a statement from the U.S. Fish and Wildlife Service (USFWS) that the plan might not meet the criteria for delisting wolves. Wildlife officials from Montana and Idaho urged Wyoming to avoid this classification, claiming that it would delay action to delist the wolf.

Further events showed the reluctance of Wyoming to undertake reasonable management of this controversial animal. State lawmakers in recent years passed three bills challenging the authority of the federal government to define the criteria for managing wolves.[2] In addition, Dave Moody, large predator coordinator for Wyoming Game and Fish Department (WGFD), was placed on administrative leave in April 2003 following critical comments he made about his agency's wolf plan.[3] Concerned about particulars of the state's plan, he said that it "does not provide long-term, adequate protection [for wolves]."[4] Although managerial control over wolves may eventually pass to the state, the willingness, ability, and underlying motives of Wyoming's wildlife management agency differ significantly from those of the USFWS, which spearheaded wolf recovery in this area. Conflict within Wyoming as well as between state and federal managers over a species that has national symbolic significance is both inevitable and problematic. Although the road ahead for wolves remains uncertain, the human drama surrounding the issue continues to unfold along well established lines.

In this chapter we examine wolf restoration in western Wyoming south of Yellowstone National Park (see figure 1.1.) since 1995. Descendants of the wolves reintroduced to Yellowstone in 1995 and 1996 have moved south and east into areas from which they were eliminated about 80 years ago. Wolves are now part of the political and social landscape. Ranchers, hunters, and others are concerned about these animals and their effects on livestock, wildlife, and issues such as local self-determination and decision making. As a practical matter, wolf management is difficult at best, and wildlife managers must make every effort to involve those who will have to coexist with wolves. Ultimately, without local and state acceptance or tolerance, wolf recovery is impossible.[5] Much has been written on the wolf restoration effort, and the wolf story has received excellent overviews by policy researchers Roberta Klein and Martin Nie, and by agency biologists and managers Doug Smith, Wayne Brewster, and Ed Bangs.[6] This chapter complements and extends their work.

We describe the natural history, populations, and management of these wolves, then analyze how the management process became politicized, and offer options to improve management. We have followed the wolf issue since the late 1970s, throughout reintroduction up to the present. We have also had numerous discussions with diverse people, attended many meetings, and read key documents about wolves over the last 25 years. Wolves first appeared in Jackson Hole in 1999 and later spread to areas farther south; this expansion enlarged the human dynamics surrounding wolf restoration. We began a formal effort to understand issues of wolf restoration in 2001, including conference calls and interviews with agency personnel, wildlife managers, livestock producers, hunters, environmentalists, wolf advocates, wolf opponents, and others. Newspaper, magazine, and journal articles provided information about wolves in general, wolf dispersal and conflict, and specifics of the social and decision-making processes that make up this case. Government documents, such as the USFWS Rocky Mountain Wolf Recovery Annual Reports, Gray Wolf Recovery Weekly Status Reports, and the Wyoming Gray Wolf Management Plan, were examined. Various Web sites were reviewed for the most current information available. Discussions with academics and colleagues who have worked on this and related issues helped us to refine our understanding of nuances in this very complex management policy problem.

Natural History, Population Dynamics, and Management History

People find wolves fascinating for different reasons. Wolves bring with them both substantive biological issues and much symbolic politics. These interact to shape management policy.

Natural History

Gray wolves (*Canis lupus*) are some of the wildest and shyest creatures in the Rockies. They generally stand 26–36 inches in height. Average length for an adult male wolf is 5–6 feet, whereas females range from 4 to 6 feet. Males typically weigh from 95 to 100 pounds and females 80 to 85 pounds. A gray coat is the most common pelage, although it may vary from black to white.[7] Wolves have blunt muzzles and carry their tails straight out when traveling, features that distinguish them from coyotes.[8] Wolves' olfactory senses are up to 100 times more sensitive than humans, and they can run up to 35–40 miles per hour.[9]

Historically, wolves ranged throughout most of North America.[10] Most people today associate these carnivores with heavily forested country, but wolves worldwide are adapted to live in a variety of landscapes and gradients of cover. They prefer low-lying and flat country and use landscape features that facilitate their movement and access to prey.[11] Wolves tend to choose areas with a low road density.[12]

Wild wolves breed at about 22 months, and typically only the dominant alpha female and male of a pack breed.[13] Female wolves come into estrus each February and March. Gestation usually lasts 63 days, and pups are born in late April or early May. Typical litter size is five pups.[14] Pups remain in the natal den for about 8 weeks, and then over the summer months the pack moves between established rendezvous sites, which are typically located in areas with abundant prey.[15] Wolves have a high reproductive rate where prey is abundant, as in Greater Yellowstone with its high densities of ungulate prey.[16] On rare occasions, subordinate females will produce pups, resulting in more than one litter per pack.

Wolves are highly adapted to hunt moose, deer, and elk. They are opportunistic and will feed on carrion and even eat vegetation if ungulates are scarce. In a Montana study, 83 percent of wolf kills were of white-tailed deer (*Odocoileus virginianus*), whereas elk (*Cervus elaphus*) and moose (*Alces alces*) composed 14 percent.[17] Wolves also consume beavers (*Castor canadensis*), coyotes (*Canis latrans*), porcupines (*Erethizon dorsatum*), ruffed grouse (*Bonasa umbellus*), snowshoe hares (*Lepus americanus*), ground squirrels (*Spermophilus* sp.), and other small mammals.[18] Wolves will also prey on domestic livestock, including cattle, sheep, llamas, horses, and goats.[19] A single wolf requires about 2.5 pounds of food per day, and their requirements double for reproductive needs. Wolves often focus their predation on the young, old, and sick animals in a prey group.[20]

Population Dynamics and Community Relations

Wolves hunt and socialize in packs.[21] Pack size typically ranges from two to eight individuals, though packs as large as 37 have been documented in Yellowstone National Park.[22] Survival of pups and recruitment into the adult population depend on predation success, lack of disease, and adequate nutrition.[23] Dispersal rates increase as average pack size and overall density increase, in effect stabilizing resident population sizes. It is possible that wolf density, rather than simply prey abundance, may be a major regulatory factor in population growth outside of human influence.[24]

Causes of wolf mortality include disease, old age, starvation, accidents, territorial disputes in which wolves kill one another, and humans. In a population of wolves mostly protected from human influence, the primary cause of mortality is territorial disputes.[25] In the Northern Rockies, the primary cause of death is humans.[26] Vehicle collisions, illegal and legal shootings, and control actions in response to livestock depredations all contribute to wolf deaths. Wolves tend to disperse over long distances.[27] Dispersing individuals are at a higher risk of death than residents, particularly when individuals move through a complex mosaic of cover and land management types, as in the Northern Rockies.[28]

Wolves exhibit territorial behavior, and a pack will defend portions of a home range against other wolves. Average territory size per pack ranges from 344 square miles in the Yellowstone region to 360 square miles in central Idaho. The density of wolf packs can lead to a tremendous variation in pack home ranges. For instance, pack territories in Montana have been documented to vary from 24 to 614 square miles.[29]

When considering the ecological role of predators, the impacts on an ecological system are often hidden from the view of a casual observer. Through a top-down trophic cascade, wolves alter the numbers and behavior of their prey, and this in turn alters other aspects of the system. For example, wolf predation changes elk behavior, including browsing patterns, foraging behavior, and overall movement patterns. The change in elk feeding habits may change the spatial composition of aspen regeneration.[30] The ecological implications of the presence of carnivores in this case are indirect but significant.

Management History

Wolf management is highly controversial. Historically, wolves have nearly always been seen as a problem. Livestock newly imported to the West in the 1870s were believed to be at great risk, so wolves were hunted, trapped, and poisoned in subsequent decades. Cattlemen and sheep producers tried hard to eradicate wolves. A 1905 article in the *Pinedale Roundup* reported that "sights now to be witnessed on the range are such as to take all the heart away from the sturdy ranchman who has worked for the upbuilding of this country. The damnable terrorists of the timber—the wolves—are sweeping down on the range and playing havoc with the cattle."[31] With pressure from livestock producers, state and federal agencies entered the wolf eradication business: 20,819 wolves were killed in Wyoming between 1896 and 1907.[32] The Wyoming wolf population became functionally extinct when the last wolf was killed in Yellowstone National Park in 1924, and by the 1930s wolves had been virtually exterminated from the Rocky Mountain West.[33] The last living wild wolf in the Yellowstone region was killed in 1944 on the Wind River Reservation in Wyoming.[34] Wolves were absent from the U.S. Northern Rockies until 1986, when a den was found in northwestern Montana.[35]

The Rocky Mountain wolf was listed as an endangered species under the Endangered Species Act in 1973.[36] This law called for the federal government to do everything in its power to reestablish sustainable wolf populations. In 1974 a study team was assigned to develop a wolf recovery plan, which was approved in 1980.[37] Shortly thereafter in 1982, the Endangered Species Act was amended to include section 10(j), which stated that animals reintroduced to historical range could be classified as "experimental."[38] This classification gave wildlife managers flexibility in management activities, in effect allowing the removal of nuisance animals that would otherwise be protected as an endangered species. In 1987 the Department of the Interior recommended that wolves be returned to the Yellowstone ecosystem, and in 1988 a study was formalized.[39]

In response to strong public opposition to possible reintroduction, in 1990 the Department of the Interior developed a Wolf Management Committee of state and federal officials and representatives of conservation groups, livestock groups, and hunting groups to draft a new compromise wolf recovery plan.[40] After many public scoping meetings, this advisory committee recommended that wolves be reintroduced into Yellowstone National Park and possibly central Idaho. This plan recommended that management become the responsibility of the states. The U.S. Congress rejected this proposal and directed the USFWS, in coordination with the National Park Service, to develop an Environmental Impact Statement for wolves. The USFWS produced a draft environmental impact statement on gray wolf recovery in the Northern Rockies in 1993. This document offered alternatives for wolf reintroduction, ranging from no reintroduction to reintroduction of wolves with varying degrees of protection.[41] This draft statement received more than 160,000 public comments, the majority of which showed support for wolf restoration. In 1994 the USFWS final environmental impact statement was approved, and the selected alternative was to reintroduce wolves. Opponents of the reintroduction challenged this decision in court. Ultimately, a federal appeals court in Denver ruled that wolves would be reintroduced and deemed a "non-essential, experimental" population.[42]

Wolf recovery in the Northern Rockies is centered on three main areas—the Greater Yellowstone Recovery Area, the Central Idaho Recovery Area, and the Northwest Recovery Area in Montana. Wolves were reintroduced to Yellowstone and central Idaho in 1995 and 1996, through cooperation with the National Park Service and the Canadian government. Wolves captured in Alberta and British Columbia were translocated to Yellowstone. A total of 31 wolves was released into Yellowstone National Park in three releases, and these served as the source for future wolf dispersal to the south and east of the park.[43] In 1999 a member of the Jackson Hole pair gave birth to the first litter of wolves to be born in the Jackson Hole area in 60 years.[44] As of the beginning of 2004, 13 separate packs had formed outside of Yellowstone National Park in Wyoming, consisting of a total of 82 individual wolves. The overall wolf population in the Northern Rocky Mountains has increased exponentially since reintroduction, though recently the rate of population growth has slowed in some areas. The wolf population in the Northern Rockies grew by around 15 percent in 2003, down from 18 percent growth in 2002, and 28 percent growth in 2001. As of late December 2003, the Northern Rockies held 761 wolves, including 51 breeding pairs. The Greater Yellowstone Recovery Area held 301 of these wolves, including 16 breeding pairs.[45] The USFWS considered 30 reproducing packs over the three-state area for three successive years sufficient for the removal of Endangered Species Act protections for the wolf. This population target was first achieved in 2000, but removal from Endangered Species Act protection remains contested and has been postponed.[46] As part of this process, on March 18, 2003, the gray wolf was "downlisted" from endangered to threatened status in Idaho, Montana, and Wyoming. The "nonessential, experimental" designation still applies.[47] Meanwhile, increases in wolf numbers and expansion of the areas occupied have led to increased conflicts with traditional land uses.

A major step toward complete delisting of the wolf came in December 2003. The three management plans from Montana, Idaho, and

Wyoming were reviewed by 11 wolf experts selected by the USFWS.[48] The primary question presented to the reviewers was, "Collectively, will the three state wolf management plans conserve a recovered wolf population should the Endangered Species Act protections be removed?" Although their report generally found that the plans were adequate to meet the stated recovery goal of 30 breeding pairs in the three-state area, members of the review panel voiced concerns. Paramount among these was whether the Wyoming plan to classify wolves as predators outside a zone including national parks and adjacent wilderness areas, thereby allowing unregulated take, was reasonable under the restoration goal. Another major concern was that all three state plans rely extensively on federal funding for wolf management in the future. This funding has not been established or secured, nor are its prospects known.[49]

In early January 2004, the USFWS rejected Wyoming's wolf management plan, primarily because of its hard-line position in classifying wolves as predators. Since delisting is tied to USFWS acceptance of the collective plans of all three states, delisting has been put on hold until the issues with the Wyoming plan have been resolved. Steve Williams, director of the USFWS, said that the predator status would allow for unregulated killing and inadequate monitoring and would not ensure viable population levels. Meanwhile, Marvin Applequist, executive vice president of the Wyoming Farm Bureau, voiced support for predator status, saying, "We don't feel that the agriculture community will have the protection, and we don't feel that the sportsmen's community is going to have the protection, either, if the feds are going to dictate the terms of the plan."[50] After considerable debate in the Wyoming legislature, the state was unable to enact legislation to change the predator status for wolves. As a result, as of early 2004 delisting remains at a standstill as Wyoming prepares to take the federal government to court over the USFWS rejection of the state's wolf plan.[51]

Politicization of Management

All aspects of wolf management are politicized. Denise Casey and Tim Clark observed in an environmental history of wolves that "wherever he goes, whatever he does, [the wolf] is burdened with a heavy load that we have laid on him—all our images of him, our dreams, our fears, our stories."[52] Wolf management is a mirror of what humans value. The management process today is full of claims and counterclaims by conservation groups, wildlife managers, hunters, livestock producers, and others. Debate occurs in the courtroom, newspapers, scientific and management arenas, and legislative sessions. There has been little meaningful dialogue among stakeholders about how to resolve value differences, largely because of limitations in the institutional system of wildlife management. People's perspectives, beliefs, and practices are central to the conflict over wolves, which focuses on matters of substance (wolf depredations, science, killing wolves, and delisting issues) and matters of process (symbolism, decision making, and power). All the while, the common interest remains elusive.

Predation on Livestock and Wildlife

The effects of wolves on livestock and wildlife are a significant point of contention in the Rockies. People's diverse value demands and the ecology of wolves create management challenges. For example, depredation on livestock results in the killing of wolves and a compensation program for livestock losses, which is itself contentious. Old West localists, especially ranchers and hunters, see wolves as having a major negative impact. Other people, especially New West environmentalists, view wolf depredation as relatively minor and acceptable. Wolf advocates consider low depredation rates a success, proof that wolves can coexist with people and livestock. To compound matters, there is no easy way to clarify the facts about wolf depredation to everyone's satisfaction, given the symbolism of the issue. Management in most cases means killing offending wolves.

Although it has been established that most wolves living near livestock do not prey on domestic animals, it does occur.[53] Generally, wolves prey on cattle less frequently than on sheep. Wolves are known to kill large numbers of sheep at any one time. For instance, at least 117 sheep were killed by wolves in central Idaho in September 2003.[54] USFWS wolf recovery coordinator, Ed Bangs, acknowledged that "sheep are very vulnerable to predators. Wolves can kill large numbers of them at a time."[55] The agency's Wyoming wolf manager, Mike Jimenez, agreed, saying, "It is not worth it with wolves in sheep country."[56]

However, wolves have killed fewer livestock than was originally anticipated in the Greater Yellowstone Recovery Area.[57] The final environmental impact statement estimated that an average of 22 cattle and 68 sheep would be killed each year.[58] Overall, the average confirmed wolf depredations of livestock between 1995 and 2002 in the Greater Yellowstone Recovery Area include just over nine cattle and approximately 41 sheep per year.[59] While predation rates vary by year and significant trends are difficult to discern, it appears that more cattle and sheep are killed each year with increasing number and distribution of wolves. Most predation has occurred on federal lands or on large ranches adjacent to federal lands.

Predation by wolves actually accounts for only a very small portion of the total annual loss of livestock. According to the National Agriculture Statistics Service, predation by coyotes, domestic dogs, mountain lions, bobcats, and other predators accounted for the loss of 3,900 cattle and 48,000 sheep in Wyoming in 2000. Livestock losses from nonpredator causes, including disease, calving, weather, poison, theft, and other unknown causes, totaled 44,100 cattle and 39,000 sheep in 2000.[60] During the same period in the entire Yellowstone Recovery area, including parts of Wyoming, Montana, and Idaho, seven cattle and 39 sheep were confirmed by the USFWS to be killed by wolves.[61] Clearly, in comparison with other sources of mortality, wolves are having little effect on livestock production. However, data about wolf depredations are less important than perceptions. For example, Terry Schram, a ranch hand for a wealthy rancher/oilman and an outspoken opponent of carnivores, has worn a t-shirt proclaiming, "Screw the bears and wolves, save the cowboy."[62]

A second matter of intense debate is predation by wolves on elk and deer. For some people, predation on native ungulates is widely accepted as part of the way nature operates. Others, however, have spearheaded a vigorous opposition, claiming that game hunting opportunities are being lost because of wolf predation. Maury Jones, an influential outfitter, said, "predators are going to destroy the wildlife in Jackson Hole."[63] Similarly, a group called Sportsmen for Fish and Wildlife distributed a flyer in November 2003 that stated, "Federal meddling, through drastically increasing predation, while eliminating feeding support is a recipe for ecological disaster. Our heritage is at stake!"[64] Opponents of wolves have a powerful voice in U.S. Senator Mike Enzi (R-WY). Loyal to local voters, Senator Enzi claimed in 2002 that "the increased threat of wolves in Wyoming is having a major impact on the state's livestock and wildlife populations." Countering the senator, a WGFD official said, "We can't say wolves are causing the herd decline. A liberal hunting season is likely the major contributor to the decline in numbers of the herd."[65] Heightening the rhetoric, hunter Robert Fanning, Jr., founder and chairman of the Montana-based group Friends of the Northern Yellowstone Elk Herd, stated, "In the time it takes to drink a cup of coffee, a wolf will run through and kill a dozen elk calves. It's a slaughterfest."[66]

Politicization of Science and Management

The science about wolves and their management is also highly politicized. Facts are difficult to get. Although some people call for more research to determine the true impact of wolves on elk, hunters call the legitimacy of government biologists into question. For example, the USFWS conducts annual monitoring of wolves in Yellowstone National Park, reporting in 2002 that 10 packs killed a total of 35 elk in 30 days. In response, Friends of the Northern Yellowstone Elk Herd showed its mistrust of federal and state wildlife biologists by conducting its own elk surveys in 2002. Their results painted a significantly different picture of the health of the elk herds than the agencies' findings. These localists claimed that cow/calf ratios (an indicator of the herd's productivity and health) showed a major decline in growth potential of the elk herd and that this was a result of wolves. Hunters' mistrust of agency biologists ran so deep that the Legislative Audit Committee in Montana voted nearly unanimously to mandate a performance review of game census techniques of the state's own agency, Montana Fish, Wildlife and Parks.[67] Science has been ineffective in tempering prevailing antiwolf sentiments.

In Wyoming, the confusion over information and its implications is similar. Although Mike Jimenez reported that all the wolf packs combined killed on average one elk per day, Mark Bruscino, a wildlife biologist for WGFD, stated that it is "hard to sort out if impacts on elk from predators are significant."[68] Groups such as the Sportsmen for Fish and Wildlife continue to contend that wolves will decimate Wyoming's elk herds, saying that "[elk] herd objectives must be increased substantially to support these voracious Canadian imports."[69] Meanwhile, elk numbers remain above the state targets.[70]

This contentious science and management debate takes place against the backdrop of history. For example, local concern over wolf impacts on elk herds is a long-standing tradition. Years ago, wolves were seen as a threat to elk herds in Wyoming.[71] In contrast, Doug Smith, wolf biologist for Yellowstone National Park, said, "By whatever means, reduction of elk is a good thing, because it restores a naturally functioning ecosystem and biodiversity."[72] Although this may be true ecologically, it does not address localist perceptions. Many local people honestly believe that wolves will bring about the demise of ungulates. They believe they have a moral obligation to protect elk, deer, and other game, as well as their own hunting opportunities. Ed Bangs put it well, "I don't know of any problems with elk herds, but I do know of problems with perceptions of elk herds."[73] Value differences underlie the conflicts over science and management.

Managing Wolves

Federal agents kill depredating wolves through legal control actions and translocations. Ranchers are reimbursed for their livestock losses through a financial compensation system developed by Defenders of Wildlife. Although intended to dampen hostility and intolerance, these actions and programs are also contentious and further politicize management.

As proven by their historical extirpation, wolves are relatively easy to kill. The Endangered Species Act's "nonessential and experimental population" designation of reintroduced wolves recognized that wolves would travel widely and cause livestock losses, a problem that could be "resolved" by killing wolves. Dave Mech, a world-renowned USFWS wolf biologist, said, "If we have learned anything, it is that the best way to ensure continued wolf survival is, ironically enough, not to protect wolves completely. If we carefully regulate wolf populations instead of overprotecting them, we can prevent a second wave of wolf hysteria, a backlash that could lead once again to persecution."[74]

Today, management largely means killing wolves. Lethal control occurs where deemed appropriate by the USFWS. On private land, there is a "one strike" policy, that is, wolves are usually destroyed after their first offense, though all control is done on a case-by-case basis. On public land, wolves get more chances to live when it has been documented that they have killed livestock. If pets are killed on private land, wolves are moved after two "strikes."[75] If pets are killed on public land, no control actions are taken. In Wyoming, livestock producers can legally shoot a wolf if it is seen chasing livestock on private land. Between 1995 and 2003, a total of 42 wolves were physically relocated and 95 wolves were killed in the Greater Yellowstone Recovery Area.[76] At times, the control actions undertaken by federal managers have elicited strong reactions from those concerned about wolf welfare.[77] Although wolf management in practice amounts to killing wolves, management of livestock generally stays the same. Other management options do exist, though, as discussed later in this chapter.

The compensation program was developed in the late 1980s to build tolerance for wolves in the livestock production community. Ed Bangs, wolf recovery coordinator for the USFWS, supported the feeling of the livestock community, saying, "If you're the guy getting whacked, it is a big deal."[78] The program's goal, as stated by Defenders of Wildlife, is "to shift economic responsibility for wolf recovery away from the individual rancher and toward the millions of people who want to see wolf populations restored. When ranchers alone are forced to bear the cost of wolf recovery, it creates animosity and ill will toward the wolf. Such negative attitudes can result in illegal killing." Payments in western Wyoming from 1987 through January of 2003 totaled $70,441.47, compensating for 52 cattle, 130 sheep, 13 dogs, and 5 horses. Ed Bangs holds this program in high regard, stating, "This program should be a model for others who want positive solutions for complex environmental issues. The livestock compensation program certainly made wolves tolerable to livestock producers . . . and has made wolf recovery more easily attainable."[79]

However, the program has had its critics. First, compensation requires that federal agents confirm that wolves killed the livestock. Confirmation currently is a joint responsibility of the Department of Agriculture's Animal and Plant Health Inspection Service Wildlife Services program and the USFWS. In frustration with this program, one livestock producer said, "It is virtually impossible to get them to confirm a kill."[80] Second, ranchers claim that livestock death is not the only impact of wolves—for example, sheep are stressed and do not gain weight as fast as they would without wolves present. Third, although Defenders of Wildlife pays 100 percent of the market value of lost livestock up to $2,000, some claim that this is problematic, that some of the livestock killed by wolves have a market value well above the existing reimbursement cap, and that the value of the animal at sale is the amount of money that should be paid.[81] Fourth, ranchers claim that it reflects badly on their skill and respect when they lose livestock to wolves.

Clearly, more than money is required to ameliorate the losses of ranchers. Mike Jimenez, who is in the field almost daily, says the compensation program does not have the flexibility to work with producers on a one-to-one basis. In effect, the compensation program rewards ranchers for raising livestock for wolves without examining how their husbandry practices might be changed to minimize losses in the first place. The compensation program is slated to end when wolves are removed from Endangered Species Act protection and when management is transferred to the State of Wyoming.[82] The program is under review now by Defenders of Wildlife, but the future of compensation is in doubt. According to Kim Barber of the U.S. Forest Service and Mark Bruscino of the WGFD, Wyoming's grizzly bear damage program has been very difficult to manage.[83] A wolf compensation program at the state level may be even more difficult. Interestingly, the predator status for wolves promoted by the Wyoming Game and Fish Commission denies livestock producers the opportunity for compensation, and it may provide an incentive for the killing of wolves, whether they prey on livestock or not.

Delisting and Future Management

Once wolves are declared recovered, management will be turned over to the State of Wyoming and the WGFD. Ed Bangs stated in 1997, "I truly believe the best place for wolf management is with the state game and fish agencies. They manage the wolf food."[84] But given Wyoming's history and its present animosity toward wolves, many people question if the state has the commitment and capability to manage wolves well. State management may well translate into killing more wolves. This possibility has further politicized the management process.

Wyoming is having a difficult time sorting out its role, goals, and resources with regard to wolf management. The first question is what role the department should take in wolf management. When initially faced with the prospect of assuming wolf management in 1996, the WGFD conducted a public opinion survey.[85] Results showed that the rural public did not want wolves and that WGFD should not be responsible for managing them. In addition, the Wyoming Game and Fish Commission has consistently made its states' rights position known. In 1997 the commission opted not to participate in the wolf issues that were unfolding at the time, reengaging only when it was apparent that the delisting process could not occur without explicit state involvement. Until that time, the WGFD had adopted a "wait and see mode at the direction of the commission."[86] Recently, WGFD has released numerous publications that outline its position and formula for success in this complex case. Officials assert, "The State of Wyoming didn't ask for wolves, and Wyoming's people are fairly split on the opinion to have wolves here. Regardless, we have them now and as an agency, we will manage wolves for specific population objectives in balance with other state wildlife management objectives."[87] Although this may sound straightforward, there are many embedded issues at play in these comments, and it is clear that achieving "balance" will be a highly contentious process.

Environmentalists claim that the Wyoming Game and Fish Commission is more interested in asserting its states' rights claims than in genuinely trying to work with legitimate interests to address wolf restoration in the common interest. On the other hand, others, including outfitter Maury Jones, claim that Wyoming is bending too far in favor of wolves. He says the "Game and Fish Department is working hand-in-glove with the enviro-wackos."[88] Without a clear strategy to help people work together, the WGFD will continue to be marginalized while being stuck between divergent claims.

The second problem for future management by Wyoming is the clarification of goals. If the state assumes management, should wolves be classified as trophy big game animals with regulated hunting, or should wolves be listed as predators with unregulated killing? It remains to be seen whether the three state plans will be sufficient for delisting. Once the states take over management, hunting and control can become easier to implement. Some see this increased local control as exactly what is needed, but others feel great trepidation about the ability of the WGFD to uphold best management standards for this species. Given Wyoming's politics and the department's history, it seems likely that wolves will be heavily controlled under state management in the future.

The third consideration is resources. Money is needed to plan and manage. In 2002 the Wyoming Game and Fish Commission requested federal funds before it would even consider writing a wolf management plan, and it held out until the USFWS provided $150,000.[89]

Today WGFD is funded almost exclusively from hunting and fishing license fees and federal taxes on the purchase of hunting and fishing equipment.[90] Idaho, Montana, and Wyoming wildlife managers frequently voice concern about how wolf management will be funded, stating that current sources of revenue will be insufficient. Wyoming has estimated that it will cost approximately $1 million dollars annually.[91] This increase in required resources may dilute the effectiveness of an already financially strapped agency.

Symbolic Politics and Power Relations

Wolf management is highly symbolic. Controversies about wolf predation on livestock and wildlife, science, control killing, compensation programs, and transition to state management are only the tip of the iceberg. At the heart of wolf management is a struggle about who benefits and who pays, about who gets more power—or respect, or money—and who gets less, and so on. The power struggle between state and federal governments continues, typically played out as a contest between local and national interests, and often played out symbolically. These dynamics will always be with us, regardless of who manages wolves, but they do not condemn us to a future that will be as contentious as the past has been. Clarifying and securing people's common interests in wolf management is certainly difficult, but it can be facilitated to achieve better outcomes.

The wolf drama is but one battle in the difficult transition from historic land uses and beliefs championed by Old West localists, the State of Wyoming, and the Wyoming Game and Fish Commission to modern practices that are bolstered by different beliefs and perspectives. Localists construe this transition as a war over the future of the West. In some ways this is an apt characterization. Wolf restoration encapsulates and symbolizes this greater, ongoing struggle in the American West over how natural resources will be used, how lives will be lived, and who gets to decide. Natural resource management, including wolf restoration, is at the symbolic forefront of a rapidly changing West.

The views of localists are clear from the claims they make and the symbols they use. They dominate the region and the state and heavily influence the governor, Wyoming Game and Fish Commission, and

WGFD. Karen Henry, president of the Wyoming Farm Bureau, said, "The issue is not wolves. . . . The issue is control of the land. This is part of a bigger agenda from the Interior Department to control the West."[92] In a comment that resonates widely with a certain constituency, Kerry White, a member of the Montana Trail Riders Association and the Montana Snowmobile Association, asked, "Do you think that they may have known the devastation the wolves were going to cause, and these environmental nuts are using wolves for a bigger agenda, to destroy hunting, ranching and ultimately move the people out?"[93] From this point of view, the federal government, anyone who does not hail from the West, and anyone who supports wolves, are all "them" in an "us vs. them" drama.

Localists feel that they are under attack. A recent *Jackson Hole Guide* editorial stated, "Bringing back the predators didn't make sense, creating countless problems and hastening the end of ranching in the valley and surrounding areas."[94] Many ranchers believe that wolves will drive them out of business, that wolves will be the final insult that will bring this lifestyle to an end. Ranchers view themselves as land stewards, but feel they have been forced to go on the defensive as their grazing practices and perspective have come under fire from environmentalists. They are convinced that wolf reintroduction was a form of "ecological imperialism" forced on them. Of course, the pressure to remain loyal to their own cultural group is a strong factor in defining their views. Respect from peers is a powerful motivation in any community and is especially strong among ranchers.

Localists are fearful that the wolf population will grow out of control and have unreasonable and unmanageable impacts on livestock and wildlife. For example, Rudy Stanko, who leases grazing rights on Bridger-Teton National Forest near Jackson, claimed that "wolves are going to get out of hand and bears are getting out of hand."[95] The rapid population growth of wolves is proof to some that the population will grow uncontrollably, cause unacceptable depredation, and ultimately threaten livestock production. They believe that "wolves must be managed like any other significant wildlife species. The looming challenge is to maintain adequate numbers in balance with other wildlife and at levels and in places that reflect the needs and interests of the people. In short, we need to manage wolves as much as we manage the rest of our wildlife."[96]

Localists feel powerless. Alan Rosenbaum, who runs cattle adjacent to Grand Teton National Park, believes that "what we say doesn't carry any weight" with the federal wildlife and land management agencies.[97] Jon and Debbie Robinett, managers of the Diamond G Ranch outside Dubois, have suffered direct and personal consequences from wolf depredations. Cattle, horses, and pets have been lost. Their daily lives are now much different, as the nearby Washakie wolf pack must be monitored daily with telemetry equipment, they must constantly be on the alert for possible depredation, and their dogs that used to run free must now be penned near their house. They feel personally affronted by these events. It is easy to understand their frustration.

The localist perspective was put into context in an article written in 2000 by Christina Cromley, who at the time was a Yale University Ph.D. student and an independent researcher with the Northern Rockies Conservation Cooperative in Jackson. She interviewed livestock producers and found that many of them spoke of the "good old days" and felt that the services provided by ranching were no longer fully appreciated by the public.[98] A fatalistic attitude underscored most of the concerns voiced by this group. After decades and often multiple generations of hard work, it is easy to understand why these people are now struggling with the prospect of a "New West." They see more and more people who have no traditional connection to the land moving in, subdividing, and voting. Wolves simply serve as a convenient focal point, a kind of shorthand for the disruptive consequences of widespread change.

These views stand in sharp contrast to the views and symbolic politics of environmentalists. To them the wolf is a symbol of wilderness unspoiled by humans. And, of course, wilderness is something to be valued, left alone, and even revered. For example, an article in the *New York Times* noted that wolves "epitomize wilderness. . . . If you have enough space and security for carnivores, you provide security and space for a lot of other species. Carnivores are canaries in the coal mine in that sense."[99] Defenders of Wildlife claimed that "we need to save wildlife like wolves for future generations of Americans."[100] Former Interior Secretary Bruce Babbitt said "wolves are a living symbol of the regard Americans have for things wild."[101]

Environmentalists have generally been better organized than localists in pursuit of their value demands. Conservation organizations, including the National Audubon Society, Defenders of Wildlife, Sierra Club, Sinapu, Predator Conservation Alliance, National Wildlife Federation, and others, have been involved in wolf advocacy for years. Their collective perspective was voiced in a recent petition to urge Secretary of the Interior, Gale Norton, to boost federal wolf protection. For Defenders of Wildlife, "restoration of these animals represents a major step in correcting earlier errors in public policy and in repairing ecological imbalances."[102] Like localists, environmentalists are enmeshed in symbolic politics, using different rhetoric in support of a different belief system.

The wolf drama is also about power relations over which level of government should manage wolves. Perhaps more than any other species, the wolf has sharpened the perennial power clash between states' rights and federalism. Livestock producer Rudy Stanko said, "What I am concerned about is everybody ignores the state's rights. Everybody follows the lead of the executives in Washington, D.C. The whole State of Wyoming is at a disservice."[103] Ranchers are not the only ones to express this view. The State of Wyoming, from the governor's office to the Game and Fish Commission, does so as well. For example, Wyoming lawmakers proposed diverting $250,000 from the state's general fund to hire a team of lawyers to sue the U.S. government on behalf of the state or private citizens who feel that the federal government has caused them undue harm through habitat or species protection measures (House Bill 300, 2003). Governor Dave Freudenthal supports this measure and sees litigation as appropriate.[104] The state's position is clear on the power relations it wants.

Other state actions further illustrate this. First, a bill passed by the 2003 Wyoming State Legislature classified wolves as "trophy game" in national forest wilderness areas that adjoin Yellowstone National Park and as "predators" in all other areas of the state as long as there are more than seven packs outside the parks and John D. Rockefeller, Jr. Memorial Parkway (House Bill 229, 2003). The Wyoming Gray Wolf Management Plan, as currently written, would manage wolves according to this classification.[105] If predator status is in place, nearly all wolves in Wyoming would be subject to unlimited hunting, since nearly all wolves range outside the park at some time each year. In addition, although wolves killed as predators would legally have to be reported, the bill does not contain language that would lead to on-the-ground enforcement of this provision. Second, the legislature considered a bill to extend funding and operations of the Wyoming Animal Damage Management Board, an agency that was created in 1999 to kill predators, mostly coyotes. Every wolf pack that uses habitat outside Yellowstone National Park would be subject to its actions. Third, the State Senate passed a bill that asserted Wyoming's "exclusive jurisdiction over wildlife."[106] This bill ordered the state's attorney general to develop a lawsuit to end all federal management of wildlife in the state. Wyoming is testing the authority and control of the federal government. The state has made similar claims and undertaken similar actions for decades in support of its states' rights power demands. This incessant and often heavy-handed demand for more power, respect, and resources makes it very problematic for the state to work with others in genuinely cooperative problem solving. Wyoming expects any "cooperation" to be on its terms and in support of its demands.

Wyoming counties express a similar outlook. County commissions have said that they want more local control, and they have used wolves and other predators to illustrate their demands. The resolutions passed by Lincoln, Sublette, and Fremont county commissions banning grizzly bears and wolves from their counties have been discussed in previous chapters. Lincoln County commissioner Stan Cooper accused the federal government, saying, "You're regulating us to death."[107] Sublette County commissioner Bill Cramer, referring to frustrations felt by locals said, "We are the American outback."[108]

Local individuals make similar demands. Some people truly believe that wolves and accompanying federal laws are impinging on their rights. Rudy Stanko asserted, "It's no one else's business how we run our predators. Maybe fish and possibly birds could be in federal jurisdiction, but definitely not animals on the ground."[109] A recent editorial in the *Jackson Hole Guide* stated, "In no way should wolves supersede the rights of ranchers or businesses or a community, but at the same time we should not allow wanton killing of the species since we have spent so much time and money bringing them to the area."[110] The debate about wolf management has become so heated that many feel justified in verbally "bashing" the federal government. Symbolic politics can be harsh.

Participants often talk right past one another. On one hand, Doug Smith, chief wolf biologist for Yellowstone National Park, said that federal managers were making an intense effort to connect with the livestock producers and rural communities outside the park, but that, regardless of their efforts, the local viewpoint remains one of skepticism.[111] His frustration over "how to reach them" is clear. On the other hand, Darlene Vaughan, a rancher from the Lander area, said, "I have nothing against the people who wanted wolves back in Yellowstone; they just were not educated about them. Wolves are not afraid of people, and they'll come right up on porches to get and kill a dog. I'm afraid for the kids, and joggers who run with dogs on a leash. People need to be careful."[112] Achieving a balance so that people and large carnivores can coexist requires all parties to join in cooperative efforts with mutual respect and a commitment to trust one another, communicate openly, and develop their problem-solving skills, rather than one side "educating" the other, disparaging the other, or overpowering the other. A suite of management options must be considered to address these conditions.

Management Options

Given the goal of restored populations of wolves and sustainable coexistence, and given the people and the cultural context, what practically might be done? Despite lingering, complex problems in the substance and process of wolf management, practical alternatives do exist for improving matters. We offer three strategic alternatives: (1) understand the context better and act on that knowledge, (2) learn and apply management lessons from wolf restoration elsewhere, and (3) use a hands-on, practice-based approach in the field. All three strategic elements emphasize working with all people, including localists, environmentalists, and agency personnel, more effectively to improve management.

Goals and Contextual Management

We recommend working much more closely with livestock producers and other localists in a variety of ways. The goal is actively to build greater acceptance and tolerance of wolves. Some ranchers are willing to work cooperatively to reduce conflicts with wolves by experimenting with new husbandry practices, for example. Some hunters are ready to assist in wolf research and management as well. Incorporating these people and their concerns to manage human-wolf conflicts could go a long way toward improving matters. A significant increase in the number of available field-based practitioners would help to make this possible. Importantly, these field staff must have an effective, problem-oriented skill set to deal with complex scenarios.

To achieve the dual goals of wolf restoration and social acceptance, we need to pay much more attention to the context. For example, Jon Robinett, a rancher from the Dunoir Valley near Dubois, said that although the USFWS had stated its concern for locals, "the feds failed to realize their intent."[113] Opportunities to integrate the needs of the ranching community are at hand. Some residents have voiced a willingness to engage in this process and want to participate.[114] Wildlife managers must adapt their approaches to engage this constituency more constructively, building on the hard work and partial success of people such as Mike Jimenez of the USFWS.

First, in addition to responding quickly to wolf predation in the field on a case-by-case basis, we should also be searching for ranchers who are interested in preventive management. One-to-one personal diplomacy may be the most effective way to work with ranchers. Working in local communities requires a significant time commitment and a genuine effort to hear and address their concerns. Although one-to-one diplomacy may seem slow moving and the gains difficult to measure, this is a critical, cumulative, and humane way to build a foundation for success. The management agencies must devote more skillful staff and resources to such efforts over the long term.

Second, strategic leadership will be needed. This should include a fundamental shift in WGFD's wildlife management approach. By focusing on the root of the wolf management problem—the people involved and how decisions are made—rather than the physical details of wolf biology, wolf restoration can be made to work for people. This will be difficult, since what is needed is "a new paradigm in how [the] agency operates."[115] It will require a demonstrated commitment and capacity to meet this new standard of management.

Third, to be realistic, those in charge of wolf restoration must fully acknowledge the human context within which restoration takes place, whether they work in the field or in the office. To date, despite substantial efforts to do this, prevailing political and social forces continue to have a large impact on the implementation of wolf restoration.

Meanwhile, perceptions of success in this regard vary. Former USFWS director Jamie Rappaport Clark said, "We used the law's protections and its flexibility to structure wolf recovery to meet the needs of the species and those of the people. This is truly an endangered species success story."[116] It may be more realistic to say that the wolf case is actually a story of a federally mandated recovery effort, backed by a national constituency, which is ultimately going to be placed in the hands of a local majority and a state government that do not want wolves or, at best, remain ambiguous about the situation. If recovery has been a biological success to date, it is still not clear how biologists and managers will work with ranchers and landowners to increase acceptance or tolerance of wolves after wolves are delisted. Since resources and support for restoration will likely diminish once the state assumes management, this is especially important. Under the traditional formula, working closely with people may simply mean killing more wolves to placate people who have problems with these animals.

Even though some people are declaring restoration a success, the program must also build adequate local acceptance from those who have to live with wolves on a daily basis. Success also depends on a state agency that is committed to long-term wolf restoration and that possesses the capabilities to meet both biological and social goals, as part of a restructured and effective institutional system of wildlife management. Without attention to these human elements, all the money, time, and work put into the program to date may not be sufficient to reach the goal of wolf restoration over the long term. As noted by Roberta Klein, a policy analyst who studied wolf restoration in the Northern Rockies, wolf management involves a mix of national and local ideologies and expectations that run the gamut of perspectives. This makes the task of finding the common interest especially complex. The precedent for intolerance has a long and well-established history. It is clear that some social change must occur to achieve broad wolf recovery and coexistence. Social change takes time and effort, and the difficulty of engaging in this task cannot be overemphasized.

Finally, the legitimate concerns of ranchers must be addressed quickly, adequately, and contextually. The success of the program will be determined largely by its social context, not by top-down, "outside" legal authority forcing change on people, and not merely by killing more wolves. Being fully contextual is the key to success.

Harvest Management Experience

We recommend harvesting the experience of wolf management from other contexts and applying successful practices in this area. The experiences and lessons learned by others can be adapted to the context of western Wyoming, allowing new and better methods to be used to address carnivore conflict and local politics. Also, existing management practices that show great promise must be identified and expanded.

First, an effort begun in Alberta a few years ago (which has unfortunately since languished because of lack of resources) could become a model for integrating people's expectations and goals. Here, a coalition of livestock growers, biologists, and managers was working collaboratively in large carnivore conservation.[117] This cooperative initiative undertaken by Timm Kaminski and others included such diverse groups as the Alberta Cattle Commission, the Alberta Fish and Wildlife Division, the Central Rockies Wolf Project, the Western Stockgrowers Association, and independent biologists. These groups shared their experiences through an open forum, including question and answer periods. The ultimate goal was to combine ranching tradition and economic needs with the specifics of conserving large carnivores. Preliminary results brought out a number of practical ideas on how to achieve these integrated management goals, including: (1) using innovative depredation-avoidance techniques on a spatial scale equal to the home ranges of social groups of wolves in all seasons, (2) accompanying monetary compensation for livestock losses with fairness and trust, and (3) getting biologists, management agencies, and environmental groups to improve their efforts to provide to ranchers a program of proactive and cost-effective assistance to help large carnivores exist.

Second, there are several good examples of partial success in practical efforts. As an improvement on the well-established compensation program, the Defenders of Wildlife Bailey Wildlife Foundation recently instituted the Proactive Carnivore Conservation Fund to encourage livestock producers to mitigate depredation events. This program provides funds for collaborative projects designed to reduce conflict through the use of electric fencing, scare devices, and livestock guarding dogs, and by finding alternative grazing areas. These and other techniques have been used with mixed success in various locations.

The most successful techniques could be further implemented in this region. It will be important to respond quickly to depredation events and to upgrade the system of confirming wolf kills of livestock.

Third, the USFWS has had some success in working with local livestock producers. Mike Jimenez is in the field on a daily basis responding to wolf and livestock operators' needs. He is an excellent model of the kind of individual who is needed. By being responsive, thorough, and committed to sustained involvement with local people, he is rebuilding trust among participants. He has found that ranchers, like many people, respond favorably to initiations of good will and respect.[118] His flexibility and adaptation to changing conditions lead to credibility with the people he deals with. Jimenez says that "civic dialogue is starting" in efforts to improve wolf management in the local context.[119] This kind of proactive approach should be expanded.

The learning-focused strategy we recommend requires three steps: first, to find and describe successful wolf conservation efforts; second, to adapt and diffuse them widely; and third, to open up new opportunities to build additional program successes. Individual and local programs must be identified, described, and carefully evaluated to see exactly how they have aided wolf conservation and coexistence. This critical appraisal to explain the formal and effective reasons for a program's success is called a "practice-based" approach to policy improvement.[120] Successful management policies are actual cases, not theoretical ones. For example, a rancher might take wolf conservation into account when planning the next year of livestock management. These cases can be used to institute "best practice" standards and can function as field-tested models to be adapted and replicated in other situations. These models shift attention away from the aggregate, overall problem or the sense of "failure," establishing in its place a constructive, positive focus that can motivate and inform actions for wolf conservation on a continuing basis.

Appraisal works best when it is both independent and continuous. Independence reduces the possibility that the appraisal process will be used to promote special interests. Appraisals are more independent when they include multiple, even competing, teams of appraisers. Some of the appraisal teams should be outside the influence of whatever organization, agency, or group commissioned the appraisal so that they have no direct stakes in the results. It is also necessary that appraisals take place on a continuous basis (quarterly, semiannually, or annually, depending on the situation). Ongoing appraisal ensures the establishment, clarity, and diffusion of good field practices. Independent and continuous appraisal obliges practitioners to use current standards and models ("best practices") as they emerge. Finally, appraisals must be conducted over a sufficient period of time to glean usable lessons. Participants must attend to appraisals over several years, and they must remain alert for compromises in the appraisal process brought about by shifts in policy, budgets, government, or other factors.

Carry Out Practice-Based Actions in the Field

We recommend that wolf restoration be shifted to a field-based, practice-based focus and away from the highly visible, seemingly intractable social conflict and symbolism that currently surrounds these issues.[121] A hands-on, cooperative, evidence-based approach will minimize carnivore conflicts while at the same time ameliorate some of the harmful political dynamics. This approach will require management actions that are flexible, responsive, and detailed enough to work on a scale appropriate to individual needs. To be successful in the long run, wolf restoration must be carried out in the field on a day-to-day basis. One-to-one credibility can be found through small-scale, practical problem solving.

First, we should directly ask livestock producers what tools they will need to be able to live with and tolerate wolves. Livestock producers are fully aware of the political ramifications of being "the bad guys," and they will tolerate large carnivores more willingly if they have access to a comprehensive suite of management tools.[122] Environmental groups, biologists, and management agencies must increase and improve their efforts to provide proactive and cost-effective assistance to ranchers.[123] This will require commitments of time, money, and people. Funding must be tiered to the level of wolf abundance. Since wolf populations and their impacts will vary over time and geography, so should the resources and tools be adaptable and available.

Second, it would be very helpful to develop site-specific management alternatives. That is, we need more refined, detailed, and spatially explicit management schemes than the generalized distinction between predator and trophy game status in Wyoming, depending on broad-scale location. Specific land use changes could be defined and applied on a very detailed scale, as opposed to continual and categorical wolf control actions. For example, specific sites crucial to wolves could be identified (to a certain extent they already are) and used to delineate areas that will be devoted primarily to the protection of wolves. Some zoning, perhaps in the spring during calving, has already been put into practice, and according to Mike Jimenez, "has worked quite well."[124] On the other side of the coin, we must acknowledge, as Jimenez says, that "there are places that cannot have wolves. . . . There are limits." Wolf advocates must accept that some places are so hostile to wolves that the species will never be allowed to return there. However, since Wyoming apparently intends to classify and manage nearly all of the state as an area where wolves can be hunted as predators, this coarse definition may not be acceptable either when considering wolf population viability. A management system that is more spatially explicit would help balance this situation.

Third, some livestock producers have indicated that depredations on livestock have been reduced in certain federal grazing allotments when they were able to move their stock out of areas of known high carnivore density or when they varied stocking rates.[125] We need to facilitate this type of proactive effort by encouraging agencies to give livestock producers the flexibility and the incentives to make on-the-ground, site-specific changes.[126] Current Forest Service regulations bind livestock producers to minimum levels of grazing allotment use to maintain their leases, enforcing a minimum use level of 90 percent of allotment capacity (determined by vegetation production) every 3 years. In addition, the number of grazing leases on Bridger-Teton National Forest is about the same as when the leases were established 100 years ago under the presumption that carnivores would not be tolerated. Yet, livestock numbers have dropped about 60 percent in the same period.[127] It would be beneficial if the national forest incorporated some flexibility into allotment regulations, allowing producers to limit stocking rates and redistribute livestock based on carnivore distribution and habits. Sheep, however, are such easy prey that their coexistence with wolves is questionable; a possible solution may be to move sheep away from wolf areas entirely. The potential also exists for the buyout of some grazing allotments. In 2003, 74,200 acres of the controversial Black Rock-Spread Creek allotment on Bridger-Teton National Forest was permanently retired because of the sheer magnitude of livestock and carnivore conflicts. Using an inclusive and proactive approach, Steve Kilpatrick of the WGFD orchestrated a partnership among livestock operators, conservation groups, and federal and state agencies that enabled the voluntary retirement of grazing activity in crucial grizzly bear and wolf habitat, effectively ending a long history of conflict. However, not all recent developments show promise. In fall 2003 Bridger-Teton National Forest proposed opening three allotments in the Wyoming Range to domestic sheep grazing. This would expand the Wyoming Range Allotment Complex. As stated earlier, when wolves and sheep inhabit the same area, conflicts are likely to occur. Developments such as this do not bode well for wolves.[128] Although no resident pack of wolves currently lives in this range, the potential for conflict exists.

The retirement of grazing allotments on federal land will require from the federal land management agencies money, labor, and most of all an understanding of the benefits.[129] Resistance by livestock producers must be anticipated since the identity and economy of the West were built on ranching. Livestock production will continue to be an important land use in Wyoming, though the number of livestock on public land is declining.[130] By minimizing contact between carnivores and livestock, depredation may decrease. Ultimately, by not integrating wildlife management and range management, federal agencies are giving livestock producers (and the public in general) mixed signals about the significance and value of these federal lands.

Conclusion

Restoring wolves and ensuring their long-term viability is relatively easy biologically, but very complex politically. The national public wants wolf restoration, but powerful segments of the local public do not. The return of wolves has raised many substantive issues, such as their predatory effects on livestock and wildlife, the use and misuse of science, and the utility of management removals and compensation programs. Since the wolf carries a huge symbolic load, it also enflames complex local and regional politics. Most importantly, the wolf drama has provoked another high-stakes round in the endless power struggle among levels of government about which should have authority and control over management of our natural resources. Symbolic and power politics, tightly entwined, override almost all other aspects of wolf restoration. In this struggle, participants behave the way they do to win symbolic victories as much as substantive gains.

At present wolves seem to be doing well. Federal agencies feel that their job is almost over and that state agencies need to assume management responsibility. The public remains divided. Overall, people have no individual or collective experience of what it takes to coexist with large carnivores over the long haul. To find an individual today who is a proponent of wolves and whose livelihood is based on livestock production is difficult.

Nevertheless, a host of practical, contextually sensitive options are available to enable people and wolves to coexist. They build on current successes and the commitment of a few dedicated individuals. Among these options is working more closely with willing ranchers, hunters, and other residents on the science and management of wolves on a case-by-case basis. Learning from other wolf management efforts and applying and adapting lessons from other areas offers another promising path. Finally, a targeted, well-timed, practice-based approach to address specific, on-the-ground problems is a proven strategy that will help as well.

In the mid- to long-term, the institutional system of wildlife management must be restructured to promote the integration of diverse interests and effective, practical problem solving. The future of large carnivores depends on a sophisticated form of participatory (individual and community-based) problem solving, learning, and adapting. Few new resources will be needed, but the effort will require skilled field people, a special kind of strategic leadership, and willing ranchers, hunters, environmentalists, and agency personnel. This is the only way to build and maintain trust, the most important component of any cooperative effort in the common interest.

References

1. "State calls wolves predators," 2002, *Jackson Hole News,* October 30, A1.

2. Wyoming Conservation Voters, Wyoming Legislative Information, Casper, WY, http://www.wyovoters.org (accessed November 29, 2004).

3. R. Huntington, 2003, "State quiets biologist who doubts wolf plan," *Jackson Hole News&Guide,* April 23, A17.

4. M. Stark, 2003, "Holes seen in Wyoming wolf plan," *Billings Gazette,* April 10, www.billingsgazette.com (accessed November 29, 2004).

5. T. W. Clark and A. Gillesberg, 2001, "Lessons from wolf restoration in Greater Yellowstone," 135–149 in V. A. Sharp, B. Norton, and S. Donnelly, eds., *Wolves and human communities: Biology, politics, and ethics,* Island Press, Washington, D.C.

6. R. Klein, 2002, "Wolf recovery in the Northern Rockies," 88–125 in R. Brunner et al., *Finding common ground: Governance and natural resources in the American West,* Yale University Press, New Haven; M. Nie, 2003, *Beyond wolves: The politics of wolf recovery and management,* University of Minnesota Press, Minneapolis; D. Smith, W. Brewster, and E. Bangs, 1999, "Wolves in the Greater Yellowstone Ecosystem: Restoration of a top carnivore in a complex management environment," in T. W. Clark et al., eds., *Carnivores in ecosystems: The Yellowstone experience,* Yale University Press, New Haven.

7. L. D. Mech, 1970, *The wolf: The ecology and behavior of an endangered species,* University of Minnesota Press, Minneapolis.

8. B. Gadd, 1995, "Wolf," *Handbook of the Canadian Rockies. Geology, plants, animals, history and recreation from Waterton/Glacier to the Yukon,* Corax Press, Jasper, Alberta, 630–632.

9. D. G. Moulton, E. H. Ashton, and J. T. Eayrs, 1960, "Studies in olfactory acuity. 4. Relative detectability of n-aliphatic acids by the dog," *Animal Behavior* 8, 117–28, cited in Mech, *The wolf;* M. H. Stenlund, 1955, *A field study of the timber wolf (*Canis lupus*) on the Superior National Forest, Minnesota,* Minn. Dept. Cons. Tech. Bull. 4, cited in Mech, *The wolf.*

10. Mech, *The wolf.*

11. K. Kunkel and D. H. Pletscher, 2001, "Winter hunting patterns of wolves in and near Glacier National Park, Montana," *Journal of Wildlife Management* 65, 520–530.

12. A. P. Wydeven, D. J. Mladenoff, T. A. Sickley, B. E. Kohn, R. P. Thiel, and J. L Hansen, 2001, "Road density as a factor in habitat selection by wolves and other carnivores in the Great Lakes Region," *Endangered Species Update* 18, 110–114.

13. Mech, *The wolf.*

14. Gadd, "Wolf."

15. Montana Department of Fish, Wildlife, and Parks, 2002, *Montana Wolf Conservation and Management Planning Document, Draft,* January, Montana Department of Fish, Wildlife and Parks, Helena.

16. Mech, *The wolf;* USFWS, 2001, *Rocky Mountain Wolf Recovery 2000 Annual Report.* http://westerngraywolf.fws.gov/annualrpt00/html/annualrpt2000.html (accessed November 29, 2004).

17. K. E. Kunkel, T. K. Ruth, D. H. Pletscher, and M. G. Hornocker, 1999, "Winter prey selection by wolves and cougars in and near Glacier National Park, Montana," *Journal of Wildlife Management* 63, 901–910.

18. D. W. Smith, 1998, *Yellowstone Wolf Project: Annual Report, 1997,* National Park Service, Yellowstone Center for Resources, Yellowstone National Park WY, YCR-

NR-98-2; D. K. Boyd, R. R. Ream, D. H. Pletscher, and M. W. Fairchild, 1994, "Prey taken by colonizing wolves and hunters in the Glacier National Park Area," *Journal of Wildlife Management* 58, 289–295.

19. Montana Department of Fish, Wildlife and Parks, *Montana Wolf Conservation and Management Planning Document, Draft.*

20. Mech, *The wolf.*

21. Mech, *The wolf.*

22. USFWS, 2002, *Rocky Mountain Wolf Recovery 2001 Annual Report,* http://westerngraywolf.fws.gov/annualrpt01/2001TAB2.pdf (accessed November 29, 2004).

23. L. D. Mech and S. G. Goyal, 1993, "Canine parvovirus effects on population change and pup survival," *Journal of Wildlife Diseases* 29, 330–333.

24. R. D. Hayes and A. S. Harested, 2000, "Demography of a recovering wolf population in the Yukon," *Canadian Journal of Zoology* 7, 36–48.

25. Mech, *The wolf.*

26. USFWS, 2000, "Proposal to reclassify and remove the gray wolf from the list of endangered and threatened wildlife in portions of the conterminous United States," *Federal Register* 65(135), 43,449–43,496.

27. D. K. Boyd and D. H. Pletscher, 1999, "Characteristics of dispersal in a colonizing wolf population in the central Rocky Mountains," *Journal of Wildlife Management* 63, 1,094–1,108.

28. D. H. Pletscher, R. R. Ream, D. K. Boyd, M. W. Fairchild, and K. E. Kunkel, 1997, "Population dynamics of a recolonizing wolf population," *Journal of Wildlife Management* 61, 459–465.

29. USFWS, 2000, "Proposal to reclassify and remove the gray wolf from the list of endangered and threatened wildlife in portions of the conterminous United States," *Federal Register* 65(135), 43,449–43,496.

30. W. J. Ripple, E. J. Larsen, R. A. Renkin, and D. W. Smith, 2001, "Trophic cascades among wolves, elk and aspen on Yellowstone National Park's northern range," *Biological Conservation* 102, 227–234; J. Berger, P. B. Stacey, L. Bellis, and M. P. Johnson, 2001, "A mammalian predator-prey imbalance: Grizzly bear and wolf extinction affect avian neotropical migrants," *Ecological Applications* 11, 947–960.

31. "Cattle hamstrung by the wolves," 1905, *Pinedale Roundup,* May 24, cited in D. Platts, 1989, *Wolf times in the Jackson Hole country: A chronicle,* Bearprint Press, Jackson, WY.

32. "Vernon Bailey on the wolf problem," 1907, *Pinedale Roundup,* February 6, cited in Platts, *Wolf times in the Jackson Hole country.*

33. WGFD, 1997, "The plan: The Wyoming Game and Fish Department's draft gray wolf recovery and management proposal 1997–2002," *Wyoming Wildlife,* Cheyenne, 16–25.

34. R. Odell, 1998, "Wolf pack might den in the Hole," *Jackson Hole News,* December 2, A1.

35. WGFD, "The plan: The Wyoming Game and Fish Department's draft gray wolf recovery and management proposal 1997–2002."

36. USFWS, 2001, *Rocky Mountain wolf recovery 2000 annual report,* http://westerngraywolf.fws.gov/annualrpt00/html/annualrpt2000.html (accessed November 29, 2004).

37. *Northern Rocky Mountain wolf recovery plan,* USFWS, Denver.

38. WGFD, "The plan: The Wyoming Game and Fish Department's draft gray wolf recovery and management proposal 1997–2002."

39. USFWS, 1994, *The reintroduction of gray wolves to Yellowstone National Park and Central Idaho, final environmental impact statement,* USFWS, Helena.

40. WGFD, "The plan: The Wyoming Game and Fish Department's draft gray wolf recovery and management proposal 1997–2002."

41. USFWS, 1993, *Draft environmental impact statement: The reintroduction of gray wolves to Yellowstone National Park and central Idaho, Summary.*

42. USFWS, *The reintroduction of gray wolves to Yellowstone National Park and Central Idaho, final environmental impact statement.*

43. USFWS, Nez Perce Tribe, National Park Service, and USDA Wildlife Services, 2003, *Rocky Mountain wolf recovery 2002 annual report,* T. Meier, ed., USFWS, Helena; R. Maughan, 2001, *Maughan's wildlife reports,* http://www.forwolves.org (accessed February 12, 2002).

44. R. Maughan, 2001, *Maughan's wildlife reports,* http://www.forwolves.org (accessed February 12, 2002).

45. USFWS, Nez Perce Tribe, National Park Service, and USDA Wildlife Services, 2004, *Rocky Mountain wolf recovery 2003 annual report,* T. Meier, ed. USFWS, Helena, http://westerngraywolf.fws.gov/annualrpt03/ (accessed November 29, 2004).

46. USFWS, *Gray wolf news, information, and recovery status reports,* http://mountain-prairie.fws.gov/wolf/index.htm (accessed November 29, 2004).

47. USFWS, 2003, *Gray wolf recovery status reports: Status of gray wolf recovery, summary of the final rule to reclassify the gray wolf March 2003,* http://mountain-prairie.fws.gov/wolf/wk04042003.htm (accessed November 29, 2004).

48. R. Huntington, 2003, "Experts give nod to state wolf plans," *Jackson Hole Daily,* December 2.

49. Huntington, "Experts give nod to state wolf plans."

50. R. Huntington, 2004, "Feds block wolf plan," *Jackson Hole News&Guide,* January 14, A1, A23.

51. J. Stanford, 2004, "Wolf bill dies: Legal fight next," *Jackson Hole Daily,* February 23, 1, 2; M. Stark, 2004, "Wyoming to sue over wolf impasse," *Billings Gazette,* March 3, http://www.billingsgazette.com/index.php?id=1&display=rednews/2004/03/03/build/wyoming/35-wolf-court.inc (accessed November 29, 2004).

52. D. Casey and T. W. Clark, 1996, *Tales of the wolf: Fifty-one stories of wolf encounters in the wild,* Homestead Publishing, Moose, WY.

53. T. W. Clark, 1999, *The natural world of Jackson Hole: An ecological primer,* Grand Teton Natural History Association, Moose, WY.

54. R. Maughan, 2003, *Maughan's wolf reports,* http://www.forwolves.org/ralph/index.html (accessed November 29, 2004).

55. "Yellowstone wolf count recently reached 218," 2002, *Bozeman Chronicle,* April 14.

56. Jimenez, USFWS, 2002, Wyoming wolf manager, pers. comm., March 11.

57. E. Bangs, J. Fontaine, M. Jimenez, T. Meier, C. Niemeyer, D. Smith, K. Murphy, D. Guernsey, L. Handegard, M. Collinge, R. Krischke, J. Shivik, C. Mack, I. Babcock, V. Asher, and D. Domenici, 2001, "Gray wolf restoration in the northwestern United States," *Endangered Species Update* 18, 147–152.

58. USFWS, 1994, *The reintroduction of gray wolves to Yellowstone National Park and Central Idaho, Final Environmental Impact Statement;* "Yellowstone wolf count recently reached 218."

59. USFWS, Nez Perce Tribe, National Park Service, and USDA Wildlife Services, *Rocky Mountain wolf recovery 2002 annual report.*

60. U.S. Department of Agriculture, 2003, National Agriculture Statistics Service, http://www.usda.gov/nass (accessed November 29, 2004).

61. USFWS, 2001, *Rocky Mountain wolf recovery 2001 annual report,* http://westerngraywolf.fws.gov/annualrpt01/2001TAB5.pdf (accessed November 29, 2004).

62. C. Cromley, 2000, "The killing of grizzly bear 209: Identifying norms of grizzly bear management," 173–220 in T. W. Clark, A. Willard, and C. Cromley, *Foundations of natural resources policy and management,* Yale University Press, New Haven.

63. M. Jones, 2002, pers. comm., March 20.

64. Sportsmen for Fish and Wildlife, 2003, "National Elk Refuge under attack!" Flyer distributed at National Elk Refuge, December.

65. "Enzi decries wolf: Senator says wolves are depleting wildlife, but facts don't back him up," 2002, *Jackson Hole News,* July 17.

66. J. Robbins, 2002, "The wolves are back, in force," *New York Times,* December 12.

67. J. Balyeat, 2002, "Sometimes crying wolf can be a good thing," *Bozeman Daily Chronicle,* February 10, A5.

68. "Gros Ventre feeders draw elk, wolves," 2002, *Jackson Hole News,* January 9, A3; M. Bruscino, wildlife biologist, WGFD, 2002, pers. comm., March 13.

69. Sportsmen for Fish and Wildlife, "National Elk Refuge under attack!"

70. R. Huntington, 2003, "State to reduce tags for Jackson bull elk," *Jackson Hole News&Guide,* April 2, A9.

71. *Jackson Hole Chronicle,* 1915, cited in Platts, Wolf times in the Jackson Hole country.

72. Robbins, "The wolves are back in force."

73. "Enzi decries wolf."

74. B. Weide and P. Tucker, 1998, "Be careful what you wish for the wolves," *High Country News,* April 13, 6.

75. M. Jimenez, 2002, USFWS Wyoming wolf manager, pers. comm., March 11.

76. USFWS, *Rocky Mountain wolf recovery 2003 annual report,* http://westerngraywolf.fws.gov/annualrpt03 (accessed November 29, 2004).

77. E. Bangs, 2001, *Letter to Fund for Animals,* http://www.r6.fws.gov/wolf/lococo.htm (accessed November 29, 2004).

78. S. McMillion, 2002, "Wolves way down on the list of sheep predators," *Bozeman Chronicle,* April 2, A1, A8.

79. Defenders of Wildlife, 2003, *Payments to ranchers from the Bailey Wildife Foundation Wolf Compensation Trust,* http://www.defenders.org/wildlife/wolf/wcstats.pdf (accessed November 29, 2004).

80. J. Robinett and D. Robinett, 2002, pers. comm., March 13.

81. M. Jimenez, USFWS Wyoming wolf manager, 2002, pers. comm., March 11.

82. USFWS, 1994, *The reintroduction of gray wolves to Yellowstone National Park and central Idaho, final environmental impact statement.*

83. K. Barber, 2002, wildlife biologist, Shoshone National Forest, pers. comm., March 13; M. Bruscino, 2002, wildlife biologist, WGFD, pers. comm., March 13.

84. D. Simpson, 1997, "Wyoming backs out of wolf management plan," *Jackson Hole Guide,* 1997, July 16, A3.

85. R. Rothwell, 2001, biologist, WGFD, pers. comm., October 25.

86. "Wolves: The Wyoming approach," 2003, *Wyoming Wildlife,* November, 34.

87. WGFD, 2003, "Wolves: The Wyoming approach," *Wyoming Wildlife,* November, 34.

88. M. Jones, 2003, Opinion: "Wolf plan doesn't follow Wyoming Law," *Jackson Hole News&Guide,* August 13, A5.

89. "Wyoming developing plan for managing wolves," 2002, *Great Falls Tribune,* June 12, M3.

90. Associated Press, 2004, "Wyoming Game and Fish Department seeks state money," *Billings Gazette,* November 11, http://www.billingsgazette.com/index.php?id=1&display=rednews/2004/11/11/build/wyoming/38-gameandfish.inc (accessed December 6, 2004).

91. WGFD, 2003, "Wolves: The Wyoming approach," *Wyoming Wildlife,* November, 34.

92. A. Halverson, 1995, "So far, wolf reintroduction survives legal challenge," *High Country News,* January 23, 3.

93. K. White, 2003, "Wolf advocates have been 100 percent wrong so far," *Bozeman Daily Chronicle,* February 9, A5.

94. T. Dewell, 2002, Editorial: "Dear Wyoming Game and Fish Commission," *Jackson Hole Guide,* October 23, A4.

95. "Gros Ventre grazer seeks eradication of wolves: Stanko raises stink over wolves, grizzlies running wild," 2002, *Jackson Hole News,* October 23, A3.

96. "Wolf-saving effort must now shift to management," 2002, *The Missoulian,* April 1, A4.

97. A. Rosenbaum, 2002, pers. comm., March 15.

98. Cromley, "The killing of grizzly bear 209: Identifying norms of grizzly bear management."

99. "Where the bears and the wolverines prey: America's wildest valley gives biologists a chance to observe carnivores in their natural element," 2002, *New York Times,* July 16, F1, F3.

100. Defenders of Wildlife, 2003, *Petition for Secretary Gale Norton,* http://www.savewolves.org (accessed November 29, 2004).

101. USFWS, 1994, *The reintroduction of gray wolves to Yellowstone National Park and Central Idaho, final environmental impact statement.*

102. R. M. Ferris, M. Shaffer, N. Fascione, H. Pellet, M. Senatore, 2003, *Places for wolves: A blueprint for restoration and long-term recovery in the lower 48 states,* Defenders of Wildlife, http://www.defenders.org/pubs/pfw01.html (accessed November 29, 2004).

103. "Gros Ventre grazer seeks eradication of wolves: Stanko raises stink over wolves, grizzlies running wild," 2002, *Jackson Hole News,* October 23, A3.

104. Wyoming Conservation Network, 2003, Weekly update on the Wyoming Legislature, in R. Maughan, 2003, Wolf reports listserve, February 25.

105. WGFD, 2003, *Final Wyoming Gray Wolf Management Plan,* July, http://gf.state.wy.us/wildlife/wildlife_management/wolf/index.asp (accessed November 29, 2004).

106. Wyoming Conservation Network, 2003, Weekly Update on the Wyoming Legislature, in R. Maughan, 2003, Wolf reports listserve, February 25.

107. "Banned in Wyoming: County officials in the Cowboy State just say 'no' to *Ursus horribilis,*" 2002, *Bozeman Chronicle,* May 4, A1.

108. B. Cramer, 2002, Sublette County Commissioners meeting, March 19.

109. "Gros Ventre grazer seeks eradication of wolves: Stanko raises stink over wolves, grizzlies running wild."

110. Dewell, "Dear Wyoming Game and Fish Commission."

111. D. Smith, 2002, comments at a workshop on "The Q Method Approach to Problem Solving in the Carnivores of the Northern Rockies," Bozeman, MT, December 18–19, sponsored by the Northern Rockies Conservation Cooperative.

112. Associated Press, "Fremont County braces for wolf battle," 2003, February 10, http://timberwolfinformation.org/info/archieve/newspapers/viewnews.cfm?ID=483 (accessed November 29, 2004).

113. J. Robinett and D. Robinett, 2002, pers. comm., March 13.

114. M. Jones, A. Rosenbaum, and A. Sommers, 2002, pers. comm., March.

115. K. Barber, wildlife biologist, 2002, Shoshone National Forest, pers. comm., March 13.

116. USFWS, 2000, "Gray wolves rebound: USFWS proposed to reclassify, delist wolves in much of the United States," press release, http://mountain-prairie.fws.gov/pressrel/00-18.htm (accessed November 29, 2004).

117. T. Kaminski, 2001, pers. comm., October 17.

118. M. Jimenez, 2001, USFWS Wyoming wolf manager, conference call with authors, October.

119. M. Jimenez, USFWS Wyoming wolf manager, 2002, pers. comm., March 11.

120. R. D. Brunner and T. W. Clark, 1997, "A practice-based approach to ecosystem management," *Conservation Biology* 11, 48–58.

121. Brunner and Clark, "A practice-based approach to ecosystem management."

122. K. Barber, 2002, wildlife biologist, Shoshone National Forest, pers. comm., March 13; M. Bruscino, 2002, wildlife biologist, WGFD, pers. comm., March 13.

123. T. Kaminski, 2002, U.S. Forest Service biologist, pers. comm., March 10.

124. M. Jimenez, 2002, USFWS Wyoming wolf manager, pers. comm., March 11.

125. A. Rosenbaum, 2002, pers. comm., March 15.

126. M. Maj, 2002, pers. comm., March 12.

127. L. Broyles, 2002, range manager, Bridger-Teton National Forest, pers. comm., March 12.

128. R. Huntington, 2003, "Feds propose grazing sheep near wildlife," *Jackson Hole News&Guide,* October 29, A13.

129. K. Barber, 2002, wildlife biologist, Shoshone National Forest, pers. comm., March 13; M. Bruscino, 2002, wildlife biologist, WGFD, pers. comm., March 13.

130. L. Broyles, 2002, range manager, Bridger-Teton National Forest, pers. comm., March 12.

Part Three:

Exploring Alternatives

Chapter **6**

Participatory Projects for Coexistence: Rebuilding Civil Society

Gregory P. McLaughlin, Steve Primm, and Murray B. Rutherford

What did I really want? A process, I think, everybody involved—ranchers, townspeople, conservationists—all taking part in that reimagining. I wanted them to each try defining the so-called land of their heart's desiring, the way they would have things if they were running the world. I wanted them to compare their versions of paradise, and notice again the ways we all want so many of the same things—like companionship in a community of people we respect, and meaningful work. —William Kittredge[1]

Western Wyoming may seem like a paradise to a good many people. In recent years, however, it has become clear that people have greatly varying "versions of paradise," to use William Kittredge's phrase. The sagebrush valleys, cold clear rivers, and snow-covered mountains have become a contested landscape. The fights are over many things at many levels: rich newcomers versus struggling old timers, federal versus local control, energy exploration versus clear skies and untouched land. One of the most visible clashes, at least in the media and public discourse, is the fight over native large carnivores, especially gray wolves, mountain lions, and grizzly bears. As the other chapters in this volume detail, each of these three species has been a focal point for controversy and conflict in the past decade.

Controversy and conflict are not necessarily problematic—healthy conflict is, of course, a natural part of politics. However, some of the struggles about large carnivores have become so chronic, so volatile, and so deadlocked that it seems almost impossible to find acceptable policies and programs to overcome them. For example, in 2002 at a public meeting in Afton, Wyoming, 400 residents packed a gymnasium to express nearly unanimous opposition to a Forest Service order setting new rules for food storage and dealing with garbage and hunting carcasses on public lands. The order was designed to reduce conflicts between bears and people, but the Forest Service encountered conflict of a different kind at the meeting in Afton and at a similar public meeting held in Lander. Kim Barber, grizzly bear specialist for the Shoshone National Forest, had the unenviable task of addressing the meeting in Lander to justify the Forest Service's new policy. "We should not have had the meeting," Barber concluded. Calling it a "hornet's nest," he endured participants' criticisms of his honesty, frustration with his agency, and distrust of the whole process. The process, he said, created heavy conflict, produced no change in the problem, and eroded faith in future attempts at collaboration.[2] In the end, the meetings in Afton and Lander led to crisis, escalating rather than diminishing conflict.

Controversies like this are not isolated events. We contend that such controversies are significant and worthy of sustained attention. It is not that large carnivores present a wholesale and imminent threat to "public health, safety, and livelihood" (as stated in the 3/12/02 official minutes of the Fremont County Commission meeting[3]), although their localized impact on some people and businesses can be acute. Rather, we think these controversies are important because the way people currently conduct such disputes is indicative of and contributes to a moribund civil society. People are failing to resolve their collective disputes or to establish an agreeable process for redressing grievances. Unfortunately, these malfunctions are not unique to the Yellowstone region; they have deep roots in our political culture.[4]

In this chapter we argue for localized, participatory projects to help resolve such conflicts and promote coexistence between people and large carnivores. Webster's Collegiate Dictionary defines participation as "the state of being related to a larger whole," stressing not only what participants do, but how they relate to one another.[5] This definition sheds some light on the large carnivore controversy. Much of the conflict about carnivores comes from fundamental problems associated with how people participate and interact, including breakdowns in their relationships with each other and their relationships with the management process. Many of the carnivore problems outlined in the previous case studies, including escalating conflicts between people and carnivores, politicization of conflicts via symbol attachment, and inability of participants and agencies to resolve conflicts, can be traced to dysfunctional processes of interaction among participants.

We begin our discussion of participatory projects by reviewing the scope of the present conflicts in western Wyoming (see figure 1.1 in chapter 1) and attempting to put them in perspective. Conflicts, by almost any definition, have increased significantly since the mid-1990s. The prospects for straightforward technical mitigation of these conflicts are constrained by a polarized sociopolitical environment that inhibits progress. Large carnivores are potent symbols in a cultural conflict of values over resource conservation and public land use in the American West, making cooperation difficult.[6] Using carnivores as symbols has magnified the actual conflicts. This conflicted context makes sweeping, large-scale improvements in carnivore management unlikely. We argue that problem-solving processes that operate at smaller scales and genuinely include citizens in conservation planning might be more successful. Such projects could make real progress at localized scales, while demonstrating an integration of values that might diminish the cultural conflict at broader scales. In addition, local participatory projects might help over time to rebuild civil society, by developing new linkages among citizens and increasing their capacity to interact constructively to form reasoned opinions about collective problems. We conclude the chapter with recommendations about how to design and implement such participatory projects.

The Contested Landscape of Large Carnivore Conservation

Although many people are frustrated with the current state of carnivore management, they remain firmly committed to the values they demand from the management process. Each participant battles to advance his or her own values, using whatever strategies of power are available, such as litigation for environmental groups, ordinances for county commissioners, petitions for other citizens, or the exercise of regulatory authority for government agencies. This aggressive pursuit of individual values, often at the expense of the values of others, increases the level of distrust and conflict. Consequently, the entire process can easily spiral into a tragic and intractable mess, where few expect things to get any better; but many feel they have no other option but to continue battling. Policy scientist Ron Brunner says of such patterns of ongoing interaction that "participants of all kinds are trapped in a complex structure that institutionalizes conflict more than it facilitates the integration or balancing of different interests into a common interest."[7]

There has been much evidence of this type of highly politicized emphasis on power in large carnivore management. When Kniffy Hamilton, supervisor of Bridger-Teton National Forest, discussed implementation of the Forest Service's food storage order, she said, "We're either going to seek collaboration to get them to accept the food order, or, if that doesn't work, [we're going to] go through with it." The Sublette County Commission, in its own bid to reclaim power at the local level, justified its ban on bears by saying that it was "fed up" with the federal government. "They're supposed to be serving us, not ramming [carnivore management laws] down our throats," said commissioner William Cramer. Meredith Taylor, the Greater Yellowstone coordinator for the Wyoming Outdoor Council, has declared that as long as wolves and grizzly bears have endangered species status, environmental groups will use the weapon of litigation to protect them.[8] Tory Taylor, a backcountry outfitter and member of the Wyoming Wildlife Federation, observed, "All groups might feel threatened about this issue because they feel like they don't have control and power. Maybe that's why they have to go to such measures. They're scared and have to act in a rash manner to get their piece of the pie. . . . People have no power over their lives, and whatever they can do to project that power over other people helps them [feel better]."[9] This exemplifies how universal frustration, distrust, anger, and a fatalistic sense of hopelessness have led many to disengage from the decision-making process, intensify conflicts, or sometimes even sabotage the entire process.[10]

Collaboration efforts arising in this environment of conflict and distrust tend to be noncommittal and may be simply symbolic gestures to satisfy regulatory, procedural requirements or buy time. Until the dysfunctional patterns of interaction are broken, no real collaboration, clarifying of common interest goals, or improvement in carnivore management will be possible.

Scope of Human-Carnivore Conflicts

To understand the politicized power struggles properly, we need to begin with a better understanding of the true scope of the conflicts. Conflicts between people and carnivores can be measured as on-the-ground ecological phenomena across time and space. Taking a long-term perspective, these conflicts were part of this region's history, but the frequency dropped dramatically by early in the 20th century as many large carnivores were extirpated. Chapter 4 outlines the range collapse of grizzlies, which resulted in the absence of the bears from most of the state except the remote backcountry in and near Yellowstone National Park. Chapter 5 describes the apparent extirpation of gray wolves in the region. Both of these species made striking comebacks, though, during the 1990s—grizzlies through sustained recovery efforts and wolves through an intensive, 2-year reintroduction effort. Changes in mountain lion populations have been more difficult to ascertain accurately, but chapter 3 describes how their numbers in western Wyoming have also probably increased in the last few decades and how there is likely to be greater concern about risks to human safety as lions are sighted more often. Overall, the recent abrupt transition from zero or few conflicts with large carnivores to dozens, or even hundreds, annually marks a significant change in the relationship between these animals and people.

Recolonization of historic range by these three species seems to be continuing. Wolves have expanded beyond the broadest delineations of the Greater Yellowstone Ecosystem, inhabiting high desert country near Farson in southwestern Wyoming.[11] Some wolves have made forays into northern Utah.[12] Grizzlies now occupy the northern end of the Wind River Range, and in 2002 a grizzly was killed near Deadman Pass in the Wyoming Range, a point closer to Utah than to Yellowstone National Park.[13] As this expansion continues, we can expect additional conflicts. At the same time, human populations are also rapidly expanding. Between 1990 and 2000, Teton, Sublette, and Lincoln counties were the fastest growing counties in Wyoming, with growth rates of 63, 22, and 15 percent, respectively.[14]

There are several ways of assessing the actual magnitude of on-the-ground conflicts between carnivores and people arising from these changes. Confirmed livestock predation is a straightforward index, but does not account for the other costs to ranchers from predation. As chapter 4 explains, death loss is only one type of damage that comes with predation. Livestock may lose weight or injure themselves in fleeing from carnivores, and they may redistribute themselves in a way that is incongruent with grazing objectives. Clearly, many ranchers perceive carnivores as serious threats to both their livestock and their traditional way of life, and finding a way to deal with these perceptions is an important step in dealing with carnivore conflicts. As Mark Bruscino, grizzly bear management officer of Wyoming Game and Fish Department (WGFD), said, there is a need to "make the situation acceptable to the guy getting impacted."[15]

Reduction of big game herds—particularly elk and moose, but also bighorn sheep—is another measure of the conflict between people and carnivores. Hunting outfitters, whose businesses depend on the availability of these game species, have mixed feelings about large carnivores. Despite inconsistent data on the actual impacts on trophy game animals, some outfitters feel that more predation by carnivores will dramatically lower game populations and thus opportunities for their clients to hunt, especially for trophy specimens.[16] For example, Maury Jones, an outfitter from Afton and leader of a 100-member outfitter association, expressed concern that "the predators are going to destroy wildlife" in the region.[17] Similarly, Wyoming Game and Fish Commissioner Les Henderson blamed mountain lions, rather than diseases spread by domestic sheep, for low numbers of bighorn sheep in the Wyoming Range. According to Henderson, the WGFD "has deliberately increased predator populations. Their objective is to decrease game populations and then blame the decrease in game on livestock grazing."[18] The anecdotal and uncertain nature of this variable makes its value dubious, but it is a popular complaint nonetheless. Though some outfitters are not concerned about large carnivores, many are currently pressuring county commissions and state and federal agencies to take a harder stand on carnivore control.

Qualitative description of changes in people's lives may yield better insight into the true scope of conflicts with carnivores than any of these other measures. Life with large carnivores is ineluctably different from life without them. Overt conflicts, such as predation on livestock or destruction of property, obviously aggrieve people. A clear example of real threats by wolves appeared in the *Sublette County Examiner* on February 6, 2003: "Dunoir rancher Jon Robinette [sic] told of the history of depredations on the Diamond G, which have persisted despite the fact that the ranch sold off a great deal of its cattle. Wolves have also killed dogs on the ranch on five occasions, coming on the ranch house's front and back porches. In one case, his wife was walking the dog to the barn to lock it up when the wolves appeared and killed it instead. In addition to cattle and dogs, wolves have killed two adult horses and a colt on the ranch as well."[19]

The preparations and precautions people must take to avoid conflicts are another form of cost. For example, the logistics of adequately securing attractants—food, livestock feed, game meat—from bears can be a time-consuming and expensive addition to a camping trip. Wolves impose similar burdens by making it inadvisable to leave horses or dogs unattended in the backcountry.[20] The presence of mountain lions in an area means that people must be more cautious about their children and pets. Apart from the logistical burdens of planning and bringing adequate equipment, carnivores may cause anxiety as people fear for their own safety as well as that of their children and animals. The number of people affected by these sorts of changes is unclear, but it may well be a majority of residents along with an unknown number of recreational visitors.

Putting Conflicts in Perspective

To address carnivore conflicts, it is important to understand the legitimate, substantive objections that people have about carnivores. However, it is also important to assess these conflicts in terms of actual social and economic impacts. This is especially true in light of the sweeping claims that have been leveled by some antipredator factions. For example, former Fremont County Commissioner Scott Luther argued that agriculture is a major part of his county's economy and that large carnivores will inevitably harm agriculture in the county and thereby cause great damage to the economy.[21] Resolutions and ordinances passed by the Fremont, Sublette, and Lincoln county governments made similar claims about threats "to public health, safety, and livelihood."[22]

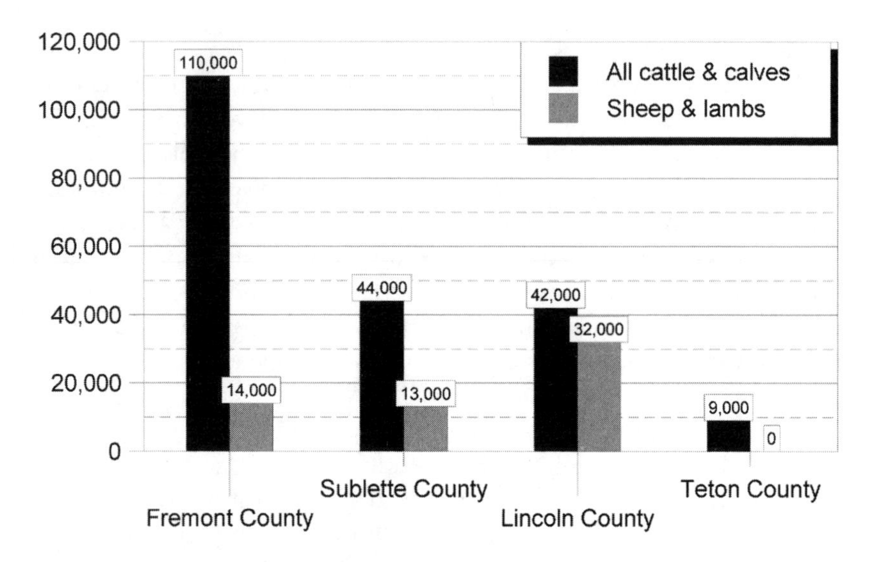

Figure 6.1
Numbers of cattle and domestic sheep in four Wyoming counties, 2001 (data source: Wyoming Agricultural Statistics Service, 2002).

It is clear that there are many cattle and domestic sheep in the southern part of Greater Yellowstone, as figure 6.1 illustrates. Fremont County, which itself is bigger than some eastern states, contains more than 100,000 cattle—more than Sublette, Lincoln, and Teton counties combined. Together, the four counties held 205,000 cattle in 2001.[23] Although no analysis has been done to evaluate the exposure of these cattle to predation risk, it seems unlikely that they are all at high risk of conflict with grizzlies, wolves, and mountain lions.

Even if large carnivores were affecting all the livestock in these counties, it is not clear that this would dramatically harm the counties' economies. Recent data gathered by economists at the Sonoran Institute cast doubt on the contention that grizzlies, wolves, and mountain lions will greatly damage local economies. Figure 6.2 shows the contribution of agriculture (including ranching and livestock rearing) as a percentage of all personal income in each of the four counties. The graph, illustrating agriculture's contribution in 1970 and in 2000, makes two key points.[24] First, agriculture is not now a major component of total personal income in these four counties. Although there

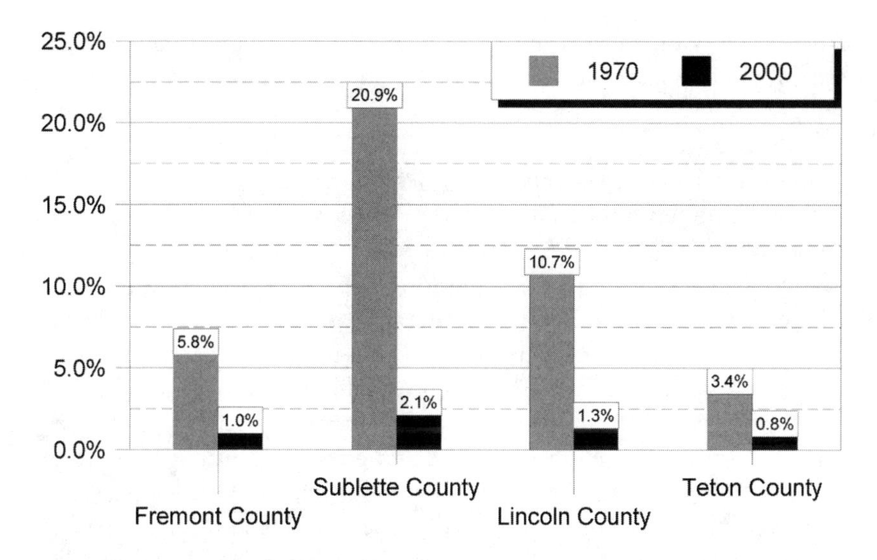

Figure 6.2
Agricultural income as percentage of total personal income in four Wyoming counties, 1970 and 2000 (data source: Rasker and Alexander 2003, using U.S. Department of Commerce statistics).

are other ways in which agriculture may be important (such as maintaining open space, paying taxes, and sustaining cultural traditions), it appears unlikely that the impacts of carnivores on agriculture will ruin local economies. Even Fremont County, with its huge numbers of cattle, does not appear to rely on agriculture for a healthy economy. The second key point is that there was obviously a time when agriculture did provide a significant portion of income, especially in Sublette and Lincoln counties.

This second point is important for understanding the social context of the region. In a relatively short time, the importance of cattle and sheep ranching has declined precipitously. There are several factors behind this decline. Low prices for beef, lamb, and wool may be one key factor. Poor livestock profit margins have forced many local people to move or find other ways to make ends meet. According to Levi Broyles, who manages grazing permits in Bridger-Teton National Forest, several longer term livestock permit holders have quit because they couldn't sustain losses: "Up to 1980 most ranching was local; now only two of the twenty allotments are local."[25] Other factors include increasing automation of certain tasks (such as hay production) that has reduced labor needs in agriculture. Also, many new, nonagricultural businesses have emerged in these counties. Finally, many new residents have entered the region in recent years, bringing nonlabor income such as retirement funds and stock dividends with them.[26] These forces are highly impersonal, diffuse, and seemingly immutable.

Ranching is central to this region's identity. That ranching is in decline—it is at least not as important as it once was economically and may be at risk of disappearing altogether—is difficult for many to accept. Large carnivores and the laws protecting them share a fragile political stage with these local woes. Of course, when national forest grazing allotments are located in carnivore habitat, there is likely to be conflict. But when family ranches face multiple economic losses, losing livestock to predation pours salt in the wounds, and carnivores become convenient scapegoats for a larger set of problems. It may be that the resurgence of large carnivores is simply the "straw that breaks the camel's back," imposing the additional cost on ranchers that could put them into insolvency. Alternatively, carnivores at least provide clear focal points, or symbols, for explaining and blaming the decline of ranching. We turn next to the role of carnivores as focal symbols.

Carnivores as Symbols and Surrogates

It should be reemphasized that effective problem solving in carnivore conflicts requires adequate information about the actual threats posed by wolves, grizzly bears, and mountain lions to livestock and human safety. Understanding the actual scope and magnitude of carnivore-human conflicts is an important first step in dealing with the problem of coexisting with carnivores. Getting a clear picture of what drives the local economy is likewise essential. In addition, though, it is vital that we have information about the broader context in order to develop a working knowledge for problem solving. One key feature of the broader context of the carnivore management problem is that grizzlies, wolves, and mountain lions have become symbolic focal points in a widespread political conflict in the Rocky Mountains.[27]

Opponents of carnivore conservation appear to be motivated by a number of grievances. Some of the opposition relates to the inherent danger that grizzlies and mountain lions can pose to humans. There is also the prospect of carnivore conservation leading to economic losses. Some opponents even claim that carnivore conservation is actually intended to curtail commercial land use and has little to do with concern for these species. Fremont County rancher Dan Ingalls stated, "the enviro groups are pushing to place these large predators everywhere because their goal is to end multiple use."[28]

It is important to understand that conflict over large carnivores reflects much larger struggles over public lands and cultural change in the West.[29] Along with the growth in both carnivore and human populations, there has been dramatic social and political change, often informally referred to as the development of the "New West." Influxes of people and wealth have created new demands in the region, which are having both positive and negative impacts on carnivores. Many of the newcomers value carnivores as symbols of wilderness, natural beauty, and quality of life. The cultural gap and increasing tension between this growing population of newer residents and the "old-timers" has been particularly obvious in Jackson. Bernie Holz, Jackson region wildlife supervisor for WGFD, highlighted this gap when he discussed the reaction of conservation-oriented residents of Jackson to the agency's grizzly bear management plan: "They aren't Wyoming people; Jackson people are in huge denial."[30]

One major reason that these struggles continue and become more entrenched is a lack of a clear, agreed-upon direction for state and federal management agencies. Land management bureaucracies such as the Forest Service are operating under a set of laws that often mandate conflicting objectives. The federal budgeting process, along with direct pressure tactics such as congressional field hearings and threats, further complicate land management. Overall, Congress has apparently expected complex laws such as the Multiple Use Sustained Yield Act and the Endangered Species Act to be integrated by the agencies on a case-by-case basis, rather than providing a clear mandate. Many observers see this as irresponsible, leading to perpetual conflict. "Rather than producing a self-correcting system, the interactions among competing groups have created a dysfunctional decision process animated by acrimony and dissatisfaction."[31]

This situation is unlikely to improve by action at the national level in the near future. A sustained, reflective, national debate on public lands, biodiversity, or large carnivores in particular seems extremely unlikely for the time being. Too many competing issues, as well as powerful political actors who would fight to keep carnivores off the national agenda, block this avenue. Powerful politicians, for example, managed to keep wolf reintroduction off the national agenda for decades. These same players are active today, employing "agenda denial" strategies such as connecting carnivores to "land grab" conspiracies and other evils.[32]

Even if grizzlies, wolves, and mountain lions did become an important issue on the national agenda, it seems unlikely that any real progress would be made. First, the odds are high that some other issue—such as health care, Social Security, or education—would quickly replace carnivores as a policy focal point. Second, reasoned debate and thoughtful problem solving seem very rare at the national level, especially about complex and esoteric issues such as carnivore conservation. Instead, the debate would likely revolve around diametrically opposed policy narratives (e.g., "wolves as noble spirit of the Rockies" vs. "wolves as evil killers") and manipulation of symbols to mobilize support.[33] Conservation advocates at this level may try to hitch carnivores to some broader agenda, such as ending public lands grazing or designating more wilderness areas, and opportunities for making real progress on the ground could be missed. Finally, the current political climate does not promise favorable outcomes for carnivore conservation through large-scale federal processes.[34]

The Fire of Conflict

In his book, *Sitting in the Fire,* Arnold Mindell explored the dynamics of group conflicts. According to Mindell, these conflicts are often replayed over time because historically destructive patterns of interaction remain unresolved in ongoing networks of relationships. These destructive patterns result from historic "public abuse" among all participants, who show symptoms such as aggression, withdrawal, silence, fear, or speaking out all the time. He defines abuse as "the unfair use of physical, psychological, or social power against others who are unable to defend themselves because they do not have equal physical, psychological, or social power," and "instead of being cloaked in silence, public abuse is out in the open where it can be witnessed by millions of people."[35] The frustrating and traumatizing impacts of these emotions burden most participants with an overwhelming sense of apathy and fatalism—almost a mass depression, in psychological terms.

Mindell's reasoning helps to explain why people engage in the carnivore process with such distrust and skepticism. It also suggests that halfhearted attempts at interaction are likely to hinder real collaborative efforts, because there is rarely a strong enough commitment from participants to form a new pattern of relating to one another. "Passivity and apathy may indicate a history of abuse," says Mindell. "Democratic countries and organizations do not function well, in part, because people who are afraid or hopeless do not represent their viewpoints." In these situations, he concludes, "a consensus is meaningless."[36]

The result of traditional carnivore management interactions is suboptimal carnivore policy. Afraid that they will lose what little access to power, wealth, or other values they have, all participants engage in historic patterns of behavior and reinforce the system of negative interaction that shrank their opportunities in the first place. In this system of conflict, even those who are peripherally involved tend to identify with the perspectives of one or more of the primary stakeholders and adopt similar ways of engaging in the process and justifying their positions. For example, many rural citizens have internalized the plight of ranchers who have experienced livestock losses. There are also cases of internal disagreement within each subgroup of participants, such that they sometimes communicate confused opinions or conflicting statements. Moreover, subgroups and the individuals who identify with them may go back and forth between contradictory viewpoints as the inner conflict plays itself out. Over time, though, people's experiences of the conflict tend to sort them into a few different shared identities—some based on direct experience and others internalized via relationships with other people. As long as these roles and identities are aligned to reinforce the historic patterns of "public abuse," improvements in the large carnivore management process are unlikely. The key to breaking these dysfunctional patterns of interaction is to get local participants working together on real, manageable, on-the-ground problems in which power and control are not such major issues and symbolic debate is minimized.

The Advantages of Participatory Approaches

Overall, considering the high costs of competing and low odds of success at the national scale, it makes more sense to pursue problem solving for carnivore conflicts in other venues and through other means. Local, genuinely participatory projects offer a viable alternative. Such projects may have the best chance of making on-the-ground progress in the high-conflict context that plagues carnivore management, while diffusing that conflict over time.

Working Together on Manageable Problems

According to Ron Brunner, more and more communities in the West are recognizing that "the old formulas for governance no longer work satisfactorily," and they are looking to find new ways to cooperate. Community-based, collaborative, or participatory approaches to problems have been recommended as a way to overcome frustration and gridlock and inspire better decisions at the local level. People are interested in participatory processes because of dissatisfaction with current decision-making processes (e.g., legislation, litigation, and administrative processes such as environmental impact statements), which seem unable to resolve contentious issues. Brunner says that the transition to local participatory processes "begins when local people realize that a pressing policy problem they experience might be solved locally."[37]

By addressing such problems locally through small-scale initiatives, the conflicts and shouting matches that overwhelm carnivore debates might be replaced with more civil and constructive discussions. Local participatory projects attempt to achieve immediate impacts on the ground, building trust among participants every time they meet their common interests. Participants learn valuable process and communication skills. In addition, the projects offer important learning opportunities. Being small scale and low profile, they provide low-stakes settings for trying out innovative ideas. Scientists, managers, and other practitioners (e.g., ranchers and outfitters) can experiment with management practices for coexistence with carnivores. Over time, successful, small-scale projects can increase political support (or lessen opposition) by serving as models for coexistence. By gaining political support, subsequent projects in other areas become easier to implement. The technical knowledge gained in one project can likewise expedite the next project. Building on successful interactions, these small experiments begin creating networks of people who are comfortable working together. As these participants interact to evaluate and apply successful experiments to new places, they may redefine the carnivore issue according to their common interest of reducing conflicts and coexisting.

Another important argument for small-scale participatory projects is that coexistence with carnivores will depend heavily on the behavior of individual people who live or work in carnivore habitat. It is unlikely that there will be sufficient political support or implementation resources for intensive monitoring and stringent enforcement of human behaviors that encourage coexistence. Instead, coexistence efforts must rely heavily on voluntary compliance and public goodwill. Voluntary compliance and informal enforcement (e.g., peer pressure) are more likely if citizens are actively involved in designing and implementing conservation programs.[38]

Fairness in Carnivore Conservation

There is also a strong fairness argument for pursuing participatory carnivore conservation. One conservative commentator has characterized current approaches to carnivore conservation as "a burden placed on the rural minority . . . by the government to satisfy the aesthetic preferences of the urban majority."[39] Although this interpretation may be overstated, it does raise important points. The Endangered Species Act makes it clear that the United States values native biota, even to the point of restoring extirpated species such as gray wolves. Yet it is also clear that this value does not trump all others; otherwise wolves, elk, bison, and other formerly widespread species would be restored throughout their historic ranges, regardless of costs or risks.

Restoring large predators and ungulates to a heavily populated state such as Ohio, for example, which has a human population density 54 times higher than that of Wyoming,[40] could be accomplished if Americans were willing to bear the tremendous costs. The majority of Americans evidently have chosen, however, largely through inaction, not to live with their full complement of native large mammals, probably because of the costs to their economies and to human safety. Yet the majority of Americans find it acceptable to impose similar costs on a tiny minority in the rural Rocky Mountain West. Sheep rancher Mary Thoman aptly summed up this dynamic: "We're paying the tab for society's project."[41] If the majority expect the minority to accept these costs, then it is only fair that rural Westerners should also have a role in designing and implementing conservation programs that is more than proportionate to their numbers. That is, if rural Westerners must bear a disproportionate burden that other Americans shirk, they should have a disproportionately bigger voice in how this burden is managed.

None of this is to say that people who are remote from current wolf, grizzly, or mountain lion ranges should have no voice in the design and implementation of conservation programs. Furthermore, there are counter arguments to be raised, particularly that the federal lands belong to all U.S. citizens. The rural West also benefits disproportionately from federal spending. Being part of the union has its benefits as well as costs.

These arguments and counter arguments will not conclusively settle the matter of participation and fairness, nor do they necessarily need to do so. Our point here is not to establish a firm legal foundation for or against local participation. Rather, we wish to canvass the persuasive claims that exist for participation in carnivore conservation. As long as the rural West is part of a larger nation that wants some large predators around, Westerners will need to abide these species. On the other hand, the larger nation must not argue that rural Westerners have some extraordinary moral obligation to conserve native biota at the expense of all other values. The upshot of these claims is that there is a mandate to conserve large predators, but people who bear disproportionate costs should have a significant role in designing how to fulfill that mandate.

Designing Participatory Processes

If we accept that local people must have a voice in order for carnivore conservation to be equitable and effective, then we need to ensure that participatory processes let this voice emerge in an accurate, well-informed way. As discussed above, the overall strategy that we advocate involves local people in addressing carnivore problems through a network of small-scale, experimental projects in areas with existing or high potential for conflicts. This experimental approach, sometimes called *prototyping,* uses small-scale, local interventions to demonstrate how problems can be solved in one location, and then adapts successful processes and solutions to other locations.[42] This approach is similar to the way in which agricultural extension agents operate soil conservation programs with farmers, but it includes both field projects and public collaboration techniques. Prototyping builds on existing information, social networks, and institutions and gives participants a legitimate reason for interacting and collaborating in carnivore management.

In the remainder of this section we discuss some of the main tasks involved in developing participatory processes that will produce reasoned, informed opinions to guide carnivore conservation.

Project Areas Scaled to Human Communities

The first step in developing a network of small-scale, participatory projects is to identify communities of people who are experiencing, or will experience, direct conflicts with carnivores. What exactly is a community of people? In his discussion of community-based initiatives, Ron Brunner described a community as a "multiple interest group," in which various participants can bring their values to the table and work to find a common interest.[43] A community, then, is simply a group of people who are affected by an issue. This "does not mean its members feel good about one another," said Brunner, only that "they are interdependent enough that they find it expedient to take each other into account."[44]

Project areas can be on the scale of watersheds, valleys, counties, Forest Service ranger districts, or other meaningful geographic or political units. The purpose of tying coexistence projects to these units is twofold. First, citizens are likely to be more engaged in participatory efforts to resolve localized conflicts with carnivores. Second, for participation to be logistically feasible, participatory processes should not place excessive burdens, such as long travel distances, on citizens. Participatory processes should also, however, include interested nonlocal people. This category could include recreationists and hunters who use local public lands but reside elsewhere. More broadly, citizens from outside the area could be represented by nongovernmental organizations with similar positions.[45]

Ideally, a local participatory project can bring members of a community together through more positive interactions than the lawsuits, executive orders, citizen petitions, and ordinances that have generally defined the large carnivore debate. Participatory projects bring people face to face, creating a variety of immediate benefits. First, the conflict becomes more humanized, making it more difficult to politicize the dialogue with exaggerated claims about either carnivores or other participants. Since people are interacting directly in a discussion about specific, on-the-ground conflicts with carnivores, symbolic debate becomes less relevant. Participants can agree at the outset on appropriate standards to ensure civility in their interactions. Participatory projects can also diffuse conflict by engaging people in a common problem-solving process. As people invest their ideas into a collective discussion, they may develop a greater sense of responsibility for the collective needs of those involved. Brunner said that this gives people "ample reason to continue to act responsibly" because "they cannot avoid the direct consequences of their actions." This encourages people to consider and work together with the real human beings sitting across from them. The many participants we have interviewed in the large carnivore conservation debate have overwhelmingly expressed a desire to return civility to the social process as a whole.

Understanding the Social Context

Another fundamental task for participatory projects is to gain a working knowledge of the area's sociocultural attributes. Valuable information includes human demographics and the makeup of the local economy. Understanding broad patterns of attitudes, beliefs, and values is key to understanding the people in the region. Of particular importance to developing successful participatory processes is identifying key individuals who are respected and influential in various subcultures of the area. Exactly who these people are will vary across settings, but it is crucial to remember that they may not always be elites in formal positions, such as county commissioners.[46] Another important facet is the area's civic institutions (e.g., the Lions Clubs, churches, and less formal organizations). Existing institutions devoted to natural resources (e.g., watershed conservation groups, conservation districts, and grazing cooperatives) are particularly significant.

Establishing wholly new efforts without first exploring existing institutions is likely to be a mistake. Experience shows that citizens who are willing and able to take a constructive role in solving conservation problems often already devote a good deal of their discretionary time to civic activities. New participatory efforts that are not integrated with existing institutions may fail to engage this limited pool of people simply because they are too busy to take on new activities. Worse, people involved with established institutions may see new efforts as competition or as a negative verdict on the usefulness of their own labors. In any event, failure to integrate with effective local institutions may doom new participatory efforts. Overlooking existing institutions, including government and state agencies, may either create active opponents or simply deprive the new initiative of the skills, leadership, and legitimacy of key individuals.[47]

A note of caution, though, is required here. Some institutions and agencies are simply not presently geared toward true community participation. For example, Bernie Holz, Jackson region wildlife supervisor for WGFD, has called collaboration "[management] by committee," giving responsibility to people who have no experience of the impacts of their decisions. Responding to a citizen-drafted grizzly bear management plan that was eventually tossed out by the department, Holz said, "the citizens groups thought they were empowered to make the decision."[48] He went on to say that community-based incentives risk creating issues that aren't issues. Other participants have found this approach extremely frustrating. Pam Lichtman of the Jackson Hole Conservation Alliance, a member of the citizens' group that drafted the rejected plan, said that it was obvious that WGFD had ignored the citizens' report and instead used the plan "they had in their back pocket the whole time."[49] Donna Wilson of the Sierra Club said that WGFD had "taken the 'public' out of public meetings."[50] Because of such experiences, participants have built up negative expectations and perceptions of some of the agencies involved in carnivore management.

A number of scholars also advise against agency domination of participatory conservation. However, it is clear that agency staff have critical roles to play, both formally and informally. Not only do managers from the state and federal agencies have extensive contact with local people, their perceptions and values may help or hinder participatory processes. Whenever possible, therefore, we suggest that managers be included in local strategies, but that these strategies be citizen-driven if they are to be effective and that the agencies not be allowed to assume roles of domination.[51]

Participatory Dialogue

The final component of our general strategy builds on the previous steps. Once a community of participants has become involved and key contextual information has been gathered, the next task is to design and implement participatory conservation strategies that produce tangible outcomes. This task is challenging, as it alters traditional roles, transcends disciplinary boundaries, and requires skills that may be in short supply. There is always a risk that participatory problem solving will end up as a new name for old, exclusionary ways of doing business. This outcome is what opponents of participatory efforts fear: replacing a transparent—if divisive and cumbersome—system of laws and procedures with small-scale deal making.[52]

People who design or are involved in participatory processes must turn to some fundamental questions from the outset. First, there must be clarity about the problem or problems they are trying to solve in a given effort. Who is it a problem for? What is their stake in solving it? How have they defined it? Most importantly, is it truly a collective problem, or a problem for one person's narrow interests? In the case of carnivore conservation, people claiming to speak for "Westerners" or "the community" may in fact be arguing only for their own particular interests. Sorting through these matters can be tiring, time-consuming, and contentious. It requires skilled facilitation, as well as a venue that is relatively free from emotional manipulation and social intimidation.[53] Some participatory efforts gloss over this important step. Too many begin with vague premises and goals and with language about "common sense," "healthy forests," or other ambiguous terms. Skipping the problem definition step in a rush to find "common ground" ultimately undermines participatory efforts.[54]

Given the widely varying perceptions surrounding carnivore conflicts, it may be worthwhile to spend the initial sessions in a process of discovery, reviewing and digesting what is known about mountain lions, grizzly bears, and wolves. The aim would be to arrive at a common understanding of the available facts and to set bounds on the limits of knowledge. In addition to the biological sciences, participants could incorporate other types of knowledge and local expertise (such as that of outfitters, artists, and Native Americans). Although this discovery phase would precede much of the discussion about value judgments, the dialogue should be structured to circle back periodically to revisit and revise understandings. Once participants establish a common foundation of knowledge, they can move on to deal with value judgments.

A second fundamental matter is the need for designers and participants to reflect on their motivations for conducting a participatory process. On the surface, people want to participate in problem-solving processes because they have detailed knowledge of the problem and have a better chance of developing solutions that are durable and reasonable. Often, however, participants really want to control the decisions that are made. If people are going to devote their time and energy to participatory problem solving, it is understandable that they would like to be assured that their preferred solutions are going to be implemented. In the broader political context, though, is this a fair demand? Taking a role in solving a collective problem is inherently political. No one else in the political arena gets an up-front guarantee that his or her preferences will prevail. Moreover, one of the primary aims of participation is to develop smarter, more reasonable solutions that satisfy a broad range of interests.[55]

Participatory processes, then, should be forums for reasoning. Solutions that emerge should be backed by "the force of a better argument," rather than by payoffs to influential interests.[56] "Better argument" can be a problematic ideal, but if participants are called upon to make their values and assumptions explicit and to provide evidence for their claims, reason can prevail.[57] If participants have carte blanche decision-making power, however, it may be difficult to follow the rule of reason. For example, the whole process may be undermined if carnivore conservation values are favored at the expense of ranching or hunting values, or vice versa. Dialogue may be cut short, and calls for explanations may be ignored. This outcome would not be much of an improvement over the status quo, since it would merely transfer power to a different party. It does not accomplish the aim of developing reasoned, durable decisions that will not be perpetually contested. Thus, many experts recommend a "firewall" between dialogue and decision making to ensure that what emerges from participatory processes is informed by better argument.[58] In other words, dialogue should focus on the underlying value conflicts and on developing mutual understanding; it should not be permitted to morph into negotiation or mediation aimed at fundamental decision making. This separation of dialogue and authority shows that participation need not be synonymous with "local control." Also, the dialogue-policy firewall may bolster the sustainability of participatory processes by avoiding the legal pitfalls of delegating statutory authority.[59]

Another advantage of the dialogue-policy firewall is that it may contribute to long-term strengthening of civil society. Civil society, according to political scientist Benjamin Barber, is "an independent domain of free social life where neither governments nor private markets are sovereign."[60] It is an amorphous concept, but most of us have experienced civil society in some form or another. Social scientists have documented the decline of "civil society" in America in recent years. People are increasingly disengaged from their fellow citizens, participating in very few associational activities.[61] We are thus left with three domains of life: the market economy, government, and the private life of home and family. The fourth domain, civil society, is vital to the resolution of collective problems. It is the proper locus of discussion, debate, and opinion formation among citizens. Government and markets are not the appropriate spheres for opinion formation because of the inherent hierarchical and coercive nature of these institutions.[62]

The absence of civil society's opinion-forming function leaves us with few viable alternatives for resolving value disputes such as those involved in carnivore conservation. Instead, we have orchestrated media campaigns and the demonization of opponents, and virtually no meaningful conversation takes place. Fortunately, civil society need not be revitalized nationwide before we can begin to restore it and benefit from it in places like western Wyoming. As environmental policy specialist Yrjo Haila pointed out, civil society is like a muscle in that it grows stronger with use.[63] Although civil society dialogues to solve carnivore conservation problems may start slowly, they can be replicated and spread and may even diffuse to nonwildlife issues as well.

Ultimately, the promise of participatory problem solving lies in its potential to find out what a spectrum of thoughtful and engaged people think about the nature of a particular problem and how we should go about solving it. Well-designed participatory processes should be an improvement over opinion polling or finding out what "the man in the street" thinks about a complex problem in the absence of information.[64] The sort of process we outline—with in-depth dialogue and a firewall between dialogue and decision-making power—may sound time-consuming and esoteric. Considering the conflict and gridlock that have characterized public land policy and wildlife conservation in recent decades, however, it is difficult to believe that the alternatives are more expedient.

Outcomes of Participatory Processes

It is unrealistic, though, to expect citizens to devote time and energy to discussions that have no bearing at all on real-world outcomes. Participatory problem solving should not be "all talk." These processes can and should focus on smaller scale, tangible outcomes, even if a firewall is maintained between dialogue and more fundamental decision making. There are many worthwhile outcomes that would improve coexistence with large carnivores, yet stop well short of "local control." These include research and monitoring, proactive conflict mitigation, and development of local conservation plans that can be readily incorporated into existing carnivore management programs. Undertaking such activities through a participatory process has two key benefits. First, these activities make substantive contributions to resolving carnivore-human conflicts, lowering the tension surrounding carnivore management efforts. Second, the participatory dimension should foster partnerships among interests who may not be traditional allies. By designing and carrying out tangible activities together, these interests can develop mutual understanding and perhaps learn to trust one another. Meanwhile, they can continue their ongoing dialogue about deeper value judgments and envisioning a future for carnivores and human communities in the local area and beyond.

Seeking Points of Entry in Decision Making

The legitimacy, relevance, and overall success of local participatory processes will depend heavily on creatively seeking out points of entry to official decision making. These points of entry could take several different forms and lead to a variety of outcomes.

First, as discussed above, participation processes should initially be aimed at designing and implementing carnivore conservation pilot projects. These could take the form of sanitation efforts, participatory monitoring efforts, or stewardship contracting to restore or improve habitat. The more "hands-on" projects would have direct local benefits and would build trust and demonstrate cooperation—powerful symbols that could lead to replication elsewhere.

Second, the discussions could lead toward the development of alternative conservation plans. After agreeing on a tentative set of value-based criteria, participants could develop plans for meeting those criteria. Representing a broad spectrum of interests and facing far fewer constraints than government agencies, it is possible that participants could uncover new ways of financing conservation efforts as well.

Third, participants from the local discussions could appoint delegates to sit on an umbrella panel to discuss broader issues of carnivores. This panel would deal with issues such as expansion of carnivore ranges and connectivity of populations. This larger panel could proceed simultaneously or could wait until the local dialogues had operated for some time.

Fourth, local processes may lead to the growth of community institutions ready to deal with new information and make timely adaptations in the conservation of carnivores and other resources. As a broader effect, these outcomes could be models for managing conservation conflicts far beyond Greater Yellowstone.

Engagement, Collaboration, and Formalization

Table 6.1 suggests how an overall strategy for participatory processes could proceed over time, considering the type of intervention contemplated, the necessary skills, the organizational resources needed to complete the work, and the accompanying message. The strategy begins with an "engagement" process, in which prototypes are directed at building trust and increasing constructive interaction among participants. As successes occur and organizational capacity and legitimacy grow, the program may become more formalized in the "collaboration" phase. Emphasizing consensus-building and joint description of alternatives, over time the program may become fully "formalized." As participatory processes become more formalized and institutionalized, the firewall between dialogue and decision making may gradually be diminished. Yearly evaluations would actually determine whether each aspect of the program should move on to a more organized, formalized level, revert to a less formal process, or terminate altogether.[65] Throughout the program, it would be necessary to develop sufficient financial and technical resources. Fortunately,

Table 6.1.
Proposed organization of a participatory carnivore management program.

Phase	I. Engagement	II. Collaboration	III. Formalization
Intervention	Small-scale prototypes (carcass removal, fencing, etc.) Participatory mapping and research	Upgraded prototypes incorporating local suggestions, skill/expertise-sharing workshops	Incentives, policies, program institutionalization, and diffusion of projects
Message	Open, inviting, humble ("We're trying to get some ideas together. What do you think?")	Mutually agreed problem articulation ("How should we be thinking about this problem?")	Mission/agenda: Formal problem redefinition and consensus ("Carnivore problem X will be met by solution Y.")
Skills	Trust-building Participatory research Collaborative learning	Consensus-building Leadership development Conflict resolution	Institution-building Long-term program implementation, grant writing
Organization type	Informal task force	Planning and problem-solving teams (informal)	Formalized decision-making body
Success indicators	Reduced conflict Increased participation	Diffusion of prototypes Increased collaboration and consensus	Low-cost programs and incentives that prevent problem resurfacing

prototyping projects are small in scale and designed to proceed without large budgets. They may also benefit from streamlining existing carnivore research and policy resources or building partnerships with organizations in the area.

Intervention

Simple, small-scale, practice-based interventions, or prototypes, can demonstrate different ways of reducing the conflict between carnivores and people.[66] Examples of prototyping activities include:

- Removal of livestock carcasses to avoid attracting predators[67]
- Development of effective fencing alternatives[68]
- Participatory mapping and research[69]

- Publishing a report showcasing views of local participants on the carnivore issue[70]
- Other interventions suggested by local participants

A notable example of a promising prototype in the Northern Rockies is a place-based and community-driven approach to the conservation of grizzly bears on private agricultural lands in the Blackfoot watershed of Montana. A major component of this effort involves a participatory GIS mapping technique that has been undertaken by Seth Wilson, a conservation biologist, and the conservation group, the Blackfoot Challenge.[71]

Wilson's mapping project involves working on a one-on-one basis with residents and ranchers to identify and map human attractants across the watershed, including beehives, bone yards (carcass dumps), calving areas, and livestock pasture locations. Montana Department of Fish, Wildlife and Parks wildlife managers are collaborating on this project and are providing data on locations of verified human-grizzly bear conflicts and locations of general grizzly bear activity. Once these data sets are assembled, Wilson will provide scientific analysis to obtain a clearer picture of where current problems are and to predict where future ones may take place. This will then enable him to work with ranchers to take concrete measures to reduce and prevent conflicts by using proven, nonlethal deterrents such as electric fencing to protect calving areas or carcass removal by Fish, Wildlife and Parks managers.

Wilson's work gives residents and ranchers an opportunity to explain their land management through mapping and discussion and reverses the traditional flow of information from expert to layperson.[72] This face-to-face interaction may help ranchers to become more invested in the carnivore conservation decision process and carnivore managers to become more aware of the needs of ranchers.

In the engagement phase, then, interventions can be introduced and supported on a small scale with willing participants, based on existing connections among participants. This also provides a unique opportunity to garner some of the knowledge of local people, leading to both better problem resolution and more participation in the process. This could lead, in the collaboration phase, to workshops emphasizing the sharing of skills and expertise. During collaboration, the participants could also evaluate and offer improvements to previous prototype projects as well as suggest new prototypes to reduce carnivore conflicts. If participants choose to formalize their efforts, prototypes could influence the development of institutionalized programs, policies, and incentive systems. This formalization and institutionalization would signal changes in how people think about and value carnivores and how the decision-making process should be carried out.

Skills

Technical support from scientists, legal scholars, and dialogue/mediation practitioners will be required to support various parts of the strategy and ensure that the work is taken seriously by decision makers. Experts will be needed to let people know the biological, political, and legal ramifications of the choices they are considering. Rather than dominating the dialogue, experts should be "on tap, not on top." In the end, no one has to sign off on the products of participatory processes if they are illegal or biologically unsound. Having experts on hand would minimize the chance of such outcomes. Skilled field workers will need to be recruited and trained for community trust-building activities such as participatory research and engagement. As the engagement phase successfully recruits participants into a dialogue about the problem, there will then need to be support for group processes, collaboration, and possible conflict resolution activities. This may simply involve a small budget for supplies and a place to meet. Finally, if the collaboration process begins to articulate formalized decisions, resources may be needed to support the building of self-sustaining, long-term solutions and organizational structures. However, formalized structures are not paramount and should be evaluated carefully for their usefulness in solving large carnivore problems, not instituted for their own existence.

Organization

Human resources are required to implement any program, and some form of organization will be necessary to ensure communication and focusing of goals. In the engagement phase, a team could simply be built around the informal association of participants sharing an interest in upgrading the current dialogue. This informal "task force" could facilitate the important steps of building connections and trust, meanwhile recruiting more participants into the discussion circle and investing them in the problem and the process. With time and success, this body could evolve into more organized, but still informal, interparticipant planning and problem-solving teams that could work together to plan more advanced prototypes and begin a more organized collaboration process. If longer term strategies gain clear consensus among participants, the organization could evolve a more formalized structure with local legitimacy and more influence with government agencies. This group could also undertake the vital task of organizing support for public officials who demonstrate skills and motivation to solve the problems.

Changing the Symbolic Meanings of Large Carnivores

We have argued throughout this analysis that large carnivore conservation needs to be released from its politically charged, symbolic associations. In other words, the "messages," myths, and meanings of large carnivores must be changed. Working together on manageable local problems is one way to decouple carnivore conservation from abstract symbolic debates. It will also be important, though, to support these projects with outreach activities that help people to think about carnivores in new ways. This is a very delicate process in which all participants need to be encouraged to evaluate their own viewpoints honestly. Like our other recommendations, outreach activities must be broadly based and participatory, rather than being imposed by one group's agenda. If a convenor is identified at an early stage in a local participatory project and is widely perceived as fair and objective, then that convenor may be the most productive outreach person to extend new messages about carnivore conservation and coexistence beyond the immediate project.

The engagement phase for changing the meanings of large carnivore conservation asks the general question, "What do you think?," and might begin by recruiting various participants (e.g., ranchers, agency staff, conservationists, and other local "experts") to provide their perceptions of the problems and potential solutions. It is important to expose the politically charged language and symbols used in discussion and to try to focus instead on interests and shared goals.[73] The viewpoints revealed during engagement might be showcased in a publication that communicates the problem less controversially and explores areas of common ground. The editors of this volume recently conducted an experimental engagement workshop of this nature, in which participants mapped out and compared their personal viewpoints on the problem of carnivore management and on potential solutions.[74] The results helped participants to understand better their own views and those of others involved.

The collaboration phase might then involve a smaller but representative group who would tackle the question, "Given this range of viewpoints, how should we be dealing with this problem?" Participants would be encouraged to discuss their values, share information, and work toward a mutually acceptable problem definition. If necessary and prudent, the formalization phase might establish organizational agendas or policy recommendations to redefine the problem formally according to shared goals.

Conclusion

Once, at a public speaking engagement, former Congressman Pat Williams (D-MT) fielded a long-winded question from a man who lived near Yellowstone National Park. The man mentioned that he was living with bison, elk, grizzly bears, and, since the 1995 reintroduction, wolves roaming across his property. As the man went on, Williams stopped him and said, "I can't tell: are you complaining or bragging?"[75]

On one hand, one might conclude that many people in Greater Yellowstone have more to brag about than to complain about. With large carnivores, we lead richer, more exciting lives. Legendary, stock-killing grizzlies have stepped out of the history books into the 21st century in the West. The wolves of Charlie Russell and Frederic Remington paintings roam the hills again. For many, it is romance incarnate. On the other hand, there are real costs associated with this romantic landscape, and they fall disproportionately on a few people. Other people who identify with this beleaguered minority take up the cause, either out of solidarity or displaced frustration, or to exploit the controversy for political gain. Large carnivores can be as powerfully negative symbols for some as they are positive symbols for others and can become a focal point for these emotions.

Participatory strategies present one of the more promising avenues for progress toward sustainable coexistence with large carnivores. We do not claim that participation is a panacea, and we fully acknowledge that in some settings and with some people, collaborative efforts may fail.[76] Also, local knowledge should be supplemented with expert knowledge, and the "practical reason" provided by participatory processes must be integrated with other governance mechanisms to ensure fairness and encourage efficiency.[77] Small-scale participatory projects, though, offer a means to work around the entrenched positions and highly charged symbolism currently associated with large carnivores in the West. It may seem odd that the alternative we propose to reduce carnivore conflicts and encourage coexistence is not so much a solution as a process. However, by engaging in collaborative problem solving at a local scale, participants can begin to reform the social process, thereby improving broader decision making about carnivores and, ultimately, rebuilding civil society.

References

1. W. Kittredge, 1996, *Who owns the West?* Mercury House, San Francisco.

2. K. Barber, 2002, pers. comm., March 13, 2002.

3. Fremont County (Wyoming) Commissioners, 2002, official minutes, March 12, http://www.fremontcounty.org.

4. See, for example, R. A. Dahl, 1982, *Dilemmas of pluralist democracy: Autonomy vs. control,* Yale University Press, New Haven.

5. *Merriam Webster's Collegiate Dictionary,* 2001, Warner, New York.

6. M. A. Nie, 2003, *Beyond wolves: The politics of wolf recovery and management,* University of Minnesota, Minneapolis.

7. R. D. Brunner, C. H. Colburn, C. M. Cromley, R. A. Klein, and E. A. Olson, eds., 2002, *Finding common ground: Governance and natural resources in the American West,* Yale University Press, New Haven, 29.

8. M. Taylor, 2002, personal interview, March 13.

9. T. Taylor, 2002, personal interview, March 13.

10. M. Taylor, 2002, personal interview, March 13; T. Taylor, 2002, personal interview, March 13.

11. C. Urbigkit, 2003, "Wolves hit Farson," *Sublette Examiner,* 1 April, http://meek.sublette.com/examiner/v3n5/v3n5s2.htm (accessed June 23, 2003).

12. B. Israelsen, 2002, "Wolf caught in Utah heads home," *Salt Lake Tribune,* December 13, http://www.sltrib.com/search (accessed November 28, 2003).

13. Interagency Grizzly Bear Study Team, 2002, press release, October 9.

14. U.S. Bureau of the Census, 2000. http://www.census.gov/prod/cen2000/phc-3-52.pdf (accessed December 1, 2004).

15. M. Bruscino, 2002, personal interview, March 12.

16. See, for example, M. Rinehart, 2002, "'Conservationists' ride public fear of

brain-wasting disease to wipe out winter feeding of elk," *Jackson Hole Guide,* August 28, A26.

17. M. Jones, 2002, personal interview, March 21.

18. C. Urbigkit, 2003, "Double sheep numbers or close allotments?" *Sublette Examiner,* May 29, http://meek.sublette.com/examiner/v3n9/v3n9s2.htm (accessed June 23, 2003).

19. C. Urbigit, 2003, "Legislators hear howling testimony," *Sublette County Examiner,* February 6.

20. For example, the popular equestrian magazine *Western Horseman* published "Survival Tips" for horse owners in wolf country, http://www.westernhorseman.com/web_extras/wolf_survival_tips.shtml (accessed November 19, 2003).

21. Scott Luther, 2002, Remarks at meeting of the Yellowstone Ecosystem Subcommittee of the Interagency Grizzly Bear Committee, November 6, http://www.fs.fed.us/r1/wildlife/igbc/Subcommittee/yes/Nov02minutes.pdf (accessed November 28, 2003).

22. Quotation from Fremont County (Wyoming) Commissioners, official minutes, March 12, 2002.

23. U.S. Department of Agriculture, National Agricultural Statistics Service, Wyoming Statistical Office, Cheyenne, http://www.nass.usda.gov/wy.

24. R. Rasker and B. Alexander, 2003, *Getting ahead in Greater Yellowstone: Making the most of our competitive advantage,* Sonoran Institute, Bozeman, MT.

25. L. Broyles, 2002, range manager, Bridger-Teton National Forest, pers. comm., March 12.

26. Rasker and Alexander, *Getting ahead in Greater Yellowstone.*

27. T. W. Clark, 2002, *The policy process: A practical guide for natural resource professionals,* Yale University Press, New Haven.

28. Ingalls and Sons, http://www.wyoag.com/Ingalls/idaho.htm (accessed 28 November 2003).

29. S. R. Kellert, M. Black, C. Reid Rush, and A. J. Bath, 1996, "Human culture and large carnivore conservation in North America," *Conservation Biology* 10, 977–990.

30. B. Holz, 2003, personal interview, March.

31. R. M. Cawley and J. Freemuth, 1997, "A critique of the multiple use framework in public lands decisionmaking," 32–44 in C. Davis, ed., *Western public lands and environmental politics,* Westview Press, Boulder, CO, 39.

32. R. W. Cobb and M. H. Ross, eds., 1997, *Cultural strategies of agenda denial: Avoidance, attack, and redefinition,* University of Kansas, Lawrence.

33. D. Stone, 1997, *Policy paradox: The art of political decision making,* Norton, New York.

34. D. Jehl, 2003, "On rules for environment, Bush sees a balance, critics a threat," *New York Times,* February 23, www.nytimes.com/2003/02/23/science/23ENVI.html (accessed February 24, 2003).

35. A. Mindell, 1995, *Sitting in the fire: Large group transformation using conflict and diversity,* Lao Tse Press, Portland, 107, 119.

36. Mindell, *Sitting in the fire: Large group transformation using conflict and diversity,* 122.

37. Brunner et al., *Finding common ground: Governance and natural resources in the American West,* 37–39.

38. G. Borrini-Feyerabend, 1997, *Beyond fences: Seeking social sustainability in conservation,* IUCN, Gland, Switzerland, http://www.iucn.org/themes/spg/Files/beyond_fences/beyond_fences.html; P. F. Starrs, 1996, "The public as agents of policy," 125–144 in *Sierra Nevada ecosystem project: Final report to Congress, v. II: Assessments and scientific basis for management options,* University of California Center for Water and Wildland Resources, Davis.

39. I. C. Sugg, 1993, "If a grizzly attacks, drop your gun," *Wall Street Journal,* June 23, A15.

40. U.S. Census Bureau, http://quickfacts.census.gov.

41. R. Huntington, 1999, "Predator troubles down for Upper Green sheep," *Jackson Hole Guide,* October 13, A5.

42. There are several examples of the application of prototypes, e.g., R. D. Brunner, 1997, "Raising standards: A prototyping strategy for undergraduate education," *Policy Sciences* 30, 167–189; R. D. Brunner and T. W. Clark, 1997, "A practice-based approach to ecosystem management," *Conservation Biology* 11, 48–58; T. W. Clark, G. N. Backhouse, and R. P. Reading, 1995, "Prototyping in endangered species recovery programmes: The eastern barred bandicoot experience," 50–62 in A. Bennett et al., *People and nature conservation: Perspectives on private land use and endangered species recovery,* Transactions of the Royal Zoological Society of New South Wales, Mosman; T. W. Clark, 1996, "Appraising threatened species recovery efforts: Practical recommendations, 1–22 in S. Stephens and S. Maxwell, eds., *Back from the brink: Refining the threatened species recovery process,* Transactions of the Royal Zoological Society of New South Wales, Mosman; S. A. Primm, 2002, "Grizzly conservation and adaptive learning," *NRCC News* (Northern Rockies Conservation Cooperative, Jackson, WY) 15, 7–8.

43. R. Brunner, C. Colburn, C. Cromley, R. Klein, and E. Olson, 2002, *Finding common ground: Governance and natural resources in the American West,* Yale University Press, New Haven, 8.

44. Brunner et al., *Finding common ground: Governance and natural resources in the American West,* 10.

45. B. Cestero, 1999, *Beyond the hundredth meeting: A field guide to collaborative conservation on the West's public lands,* Sonoran Institute, Bozeman, MT; J. M. Wondolleck and S. L. Yaffee, 2000, *Making collaboration work: Lessons from innovation in natural resource management,* Island Press, Washington, D.C.

46. Clark, *The policy process: A practical guide for natural resource professionals.*

47. Wondolleck and Yaffee, *Making collaboration work.*

48. B. Holz, 2002, personal interview, March 18.

49. P. Lichtman, 2002, personal interview, March 11.

50. Sierra Club, 2001, "Conservationists growl over Wyoming Game and Fish Department Plan to delist the grizzly bear," news release, May 1.

51. Cestero, *Beyond the hundredth meeting: A field guide to collaborative conservation on the West's public lands;* J. S. Dryzek, 2000, *Deliberative democracy and beyond: Liberals, critics, contestations,* Oxford University Press, New York.

52. Wondolleck and Yaffee, *Making collaboration work.*

53. Dryzek, *Deliberative democracy and beyond.*

54. D. Yankelovich, 1999, *The magic of dialogue: Transforming conflict into cooperation,* Simon and Schuster, New York.

55. R. D. Brunner and C. H. Colburn, 2002, "Harvesting experience," 201–247 in Brunner et al., *Finding common ground: Governance and natural resources in the American West.*

56. Dryzek, *Deliberative democracy and beyond,* 70.

57. Yankelovich, *The magic of dialogue.*

58. Dryzek, *Deliberative democracy and beyond.*

59. R. A. Kagan, 1999, "Political and legal obstacles to collaborative ecosystem planning," *Ecology Law Quarterly* 24, 871–875.

60. B. R. Barber, 1998, *A place for us: How to make society civil and democracy strong,* Hill and Wang, New York, 4.

61. Barber, *A place for us.*

62. S. A. Primm, 1996, "A pragmatic approach to grizzly bear conservation," *Conservation Biology* 10, 1026–1035.

63. Y. Haila, 1998, "Environmental problems, ecological scales and social deliberation," 65–87 in P. Glasbergen, ed., *Co-operative environmental governance: Public-private agreements as policy strategy,* Kluwer, Dordrecht, Netherlands.

64. Primm, "A pragmatic approach to grizzly bear conservation."

65. For evaluation criteria and principles of evaluation, see Clark, *The policy process: A practical guide for natural resource professionals;* and H. D. Lasswell, 1971, *A pre-view of policy sciences,* American Elsevier, New York. For a specific discussion about evaluating collaborative processes, see S. Conley and M. A. Moote, 2003, "Evaluating collaborative natural resource management," *Society & Natural Resources* 16, 371–386.

66. See endnote 42 above for references on prototypes.

67. Seth Wilson, 2004, pers. comm., January.

68. Defenders of Wildlife, 2003, press release, Wildlife agencies and environmental group team up to protect both sheep and grizzly bears, September 2, http://www.defenders.org/releases/pr2003/pr090203.html.

69. R. Margoluis and N. Salafsky, 1998, *Measures of success: Designing, managing, and monitoring conservation and development projects,* Island Press, Washington, D.C.

70. See S. R. Brown, 2003, "Workshop on large carnivore conservation," (summary report to participants), Department of Political Science, Kent State University, Kent, Ohio.

71. S. M. Wilson, 2003, *Landscape features and attractants that predispose grizzly bears to risk of conflicts with humans: A spatial and temporal analysis on privately owned agricultural land,* Ph.D. dissertation, University of Montana, Missoula.

72. S. M. Wilson, 2003, "Conservation on the edge: Lessons for recovering grizzly bears in contested landscapes," *NRCC News* 16, 3–5.

73. R. Fisher, W. Ury, and B. Patton, 1991, *Getting to yes,* 2nd ed., Penguin Books, New York.

74. Brown, "Workshop on large carnivore conservation."

75. L. Willcox, 2000, pers. comm., December 10.

76. See, for example, P. A. Walker and P. T. Hurley, 2004, "Collaboration derailed: The politics of 'community-based' natural resource management in Nevada County," *Society and Natural Resources* 17, 735–751.

77. D. Pelletier, V. Kraak, C. McCullum, U. Uusitalo, and R. Rich, 1999, "The shaping of collective values through deliberative democracy: An empirical study from New York's North Country," *Policy Sciences* 32, 103–131.

Chapter **7**

The Institutional System of Wildlife Management: Making It More Effective

Tim W. Clark and Murray B. Rutherford

A young male grizzly bear wanders south out of Yellowstone National Park in search of food or habitat and ends up in the upper Green River country northwest of Pinedale, Wyoming.[1] Along his path he encounters different landscapes and—unknown to him, of course—different human institutions, or established patterns of practices or behavior. Decisions made in the context of these institutions have led to the existence of roads, subdivisions, clear-cut forests, livestock allotments, backwoods cabins, recreational uses, motorized access, hunting, predator control, and other human activities that greatly increase the bear's chances of being killed or removed. Added to this diverse array of land uses is the staggering mix of decision makers that have some say in what happens to the land and its resources, including local, state, and federal agencies, a variety of private entities, and a growing number of nongovernmental organizations. This situation—in which the bear, going about his life, comes into contact (and often conflict) with people who are going about their lives—challenges the complex institutional system of wildlife management to act to protect the bear and the public at the same time.

As the preceding chapters show, the existing institutional arrangements for managing wildlife have failed in some ways to develop sustainable coexistence with mountain lions, grizzly bears, and wolves in the region covered by this volume (western Wyoming south of Yellowstone National Park—see figure 1.1 in chapter 1).[2] Such institutional failure in wildlife management is not unique to this region. Fred Samson and Fritz Knopf, well-known wildlife biologists, recently examined the performance of the federal agencies that dominate wildlife management on public lands in the United States (often in close partnership with state agencies and governments) and concluded that they are performing suboptimally.[3] In their view, our "archaic agencies" and "muddled missions" are simply not capable of meeting the demands of wildlife and land conservation in the 21st century. They call for better political and agency leadership, a sharper focus on mission or goals, agency structures that are less bureaucratic, and rewards for successful outcomes.

In this chapter we evaluate the present institutional system of wildlife management and propose alternatives for improving its performance. At stake in this issue is our collective ability to maintain viable wildlife populations and to prevent loss of wildlife and habitats through destructive human practices. Also at stake is public trust in government agencies and, in some cases, the democratic process itself. Finally, we are concerned about the kind of environment our children will live in—imagine a world without grizzly bears, mountain lions, or wolves. The importance of these matters calls for an honest, pragmatic appraisal of the institutional system of wildlife management, to determine what we can do to make it work better for us all. We need to ask whether the current system is the best strategy available to conserve wildlife sustainably in the public interest. If the answer is no, then we need to determine what reforms should be made. Toward this end, this chapter describes how the present wildlife management system is structured, it explains why this system operates in the ways that it does, looking at key narratives and other features in play, and it suggests how we can upgrade institutions for more effective wildlife conservation. We hope this chapter will stimulate wide discussion among professionals and the public at large.

The Capacity of the Institutional System

Sustainable coexistence between people and wildlife is the win-win, integrated outcome that most people want. Whether we can actually achieve this, however, will depend directly on the capacity of our institutions to find and implement common interest solutions for wildlife management.

Defining Institutions

Although *institution* sometimes refers to an organization, establishment, or building, we use the term more broadly to mean a "well-established and structured pattern of behavior or of relationships that is accepted as a fundamental part of a culture."[4] Policy researcher Elinor Ostrom said that institutions are "the shared concepts used by humans in repetitive situations organized by rules, norms, and strategies."[5] She emphasized how institutions provide structure and incentives for action. She called them "the rules of the game in a society or, more formally . . . the humanly derived constraints that shape human interaction."[6] Hanna Cortner and her colleagues defined institutions as "the expressions of the terms of collective human experience. Institutions reflect the ways people interact with one another and the ways they interact with their environment. Further, they are the means people use to solve social problems." Noting that the term has been used in different ways, these authors included in the definition "both formal institutions, such as administrative structures, and also informal institutions, such as customs and practices."[7]

Among the components of the wildlife management system in Wyoming are formal organizations such as the Wyoming Game and Fish Department (WGFD) and the Sierra Club, but also many other formal and informal groups and rules that guide people's behavior as well as the repeated practices that people carry out. Together, all of these patterns of behavior and actions by various groups and people, established ways of doing things, and relations among people form a complex and interconnected network that we call the "institutional system of wildlife management." This institutional system structures how wildlife is managed, how people interact, and who gets to make decisions about what (who is privileged and who is marginalized).

Social institutions grow out of the perspectives of people—their beliefs, identities, expectations, and demands. Thus, institutions arise from and are shaped by people's perspectives, practices, and cultures, but institutions in turn shape and structure how people think and act. For example, the U.S. Endangered Species Act (a congressionally authorized, formal part of the institutional arrangement) evolved out of concerns about the status of native plant and animal populations and their ecosystems. Since its enactment, this legislation has shaped and structured property relations, scientific research, and a wide range of management decisions, and it has fostered a particular view of the responsibilities of humans toward other species.

Institutions are usually related to specific social values, or desired states of affairs. For example, political institutions, including political parties and advocacy groups, have to do with how the value of power is distributed and used. Health care institutions pertain to the value of well-being and how it is produced and shared among society's members. The institutional system of wildlife management also focuses largely on well-being, in that our stewardship of the land—plants, animals, soils, air, water, minerals, and other resources—directly and profoundly affects human health, security, and welfare. As well, the wildlife management system is a direct representation of how power is distributed and used, and this, in turn, has consequences for all of the other values that people hold dear.

Democratic societies, through their institutions, promote the wide sharing among their citizens of all values—power, wealth, well-being, affection, enlightenment, skill, respect, and rectitude.[8] More tyrannical or oppressive societies, however, develop institutions that deprive their citizens of these values in various ways. For example, some societies have formal laws (or merely norms or traditions) that limit women's educational opportunities, use the military to enforce loyalty to the regime in power, or enable the widespread and destructive exploitation of natural resources. In our case, understanding how the institutional system favors or limits certain people and values is key to understanding its usefulness.

Given that institutions set and control so much of society's thinking and action, their operations should come under close scrutiny. Yet, typically, we do not know much about specific institutions and how they work; this is certainly the case for wildlife management. However, there are effective, proven methods to analyze and appraise institutions, which we can use to investigate why the institutional arrangements for wildlife management have not fully succeeded in meeting the challenge of sustainable coexistence.[9] Case studies are one powerful approach to institutional analysis, as previous chapters illustrate. From direct observation over time, we can discern patterns of decision making that show how an institution is structured and how it functions. Analytical techniques that focus on language use or discourse are also particularly useful. As policy analysts Todd Bridgman and David Barry pointed out, language "does not simply mirror the world, but instead shapes our view of it in the first place."[10] They argue that when "the importance of language in constructing policy issues is acknowledged, policy debate becomes more than just interplay between logics, or arguments—it becomes a competitive contest between discursive frameworks." The case studies in this volume clearly show that the debate over large carnivore management is just such a contest between competing discursive frameworks. Accordingly, in this chapter we focus on discourse-oriented techniques in our appraisal of the institutional system of wildlife management. Discourse analysis allows researchers or interested citizens to learn about (1) the politics associated with the discourse, (2) the effect on the polities and policies of government, (3) the effect on the institutions involved, (4) the arguments used by the critics of established institutions, and (5) the flaws in the evidence and arguments of each side. Much of this kind of information is not currently used or sought out by people in the large carnivore management process.

Institutional Structure

The *structure* of an institution or institutional system refers to the framework established for decision making, which dictates how decisions are made, by whom, and for what purposes.[11] *Function* refers to the dynamics of the individual and collective decisions that are actually made and, for wildlife management, includes whether these decisions foster coexistence with large carnivores and other common interest goals such as equity, efficiency, and inclusiveness in decision making. If the system functions suboptimally, it is possible that its structure could be modified to improve performance. The fundamental question—the basic criterion for governance in a democracy—is whether the structure and functioning of the institutional system serve the common interest.

Several structures of governance have been directly involved in bringing the mountain lion, grizzly bear, and wolf programs to their present form. In the national context, the dominant structural features are the National Environmental Policy Act and the Endangered Species Act, which have been especially important in establishing the framework under which federal agencies have made decisions about grizzly bear and wolf management. In addition, there are a host of important but more specialized structures such as recovery and management plans for species, the various committees that have produced them, and related implementation and evaluation structures. Many meetings, decisions, and actions at both federal and state levels have also structured the institutional system.

At the state level in Wyoming there are no equivalents to the Endangered Species Act and the National Environmental Policy Act. The State of Wyoming, the Wyoming Game and Fish Commission, and the WGFD are the dominant players in the institutional structure at this level. As with any institutional arrangement, this structure embodies, privileges, and advances particular beliefs, narratives, and values, which we will explore later in this chapter. The State of Wyoming has resisted the federal wildlife management structure at almost every turn for decades, preferring its own institutional approach and the value benefits that flow from it.

One window on the nature of the state's institutional structure for managing wildlife was provided by Dubois resident Robert Hoskins, who wrote an open letter to the WGFD in December 2002 questioning the adequacy of the state's draft wolf management plan.[12] He noted that the state's current approach seems to continue that described decades earlier by Ira Gabrielson, president of the Wildlife Management Institution (a consortium of professional wildlife organizations that conducted audits, among other things). In 1952 the Wildlife Management Institution audited the Wyoming Game and Fish Commission, along with the statutes under which it operated. In the report from that audit, Gabrielson wrote that "in previous studies of fish and game laws of many states, no instance has been found in which the laws give so much special consideration to livestock operators at the expense of the fish and game resources as is found in Wyoming. . . . These laws give consideration to a minority group far beyond that found necessary or desirable in any other state studies. . . . It is obvious that in some cases the earmarking of Fish and Game funds for these purposes by legislative action has so many undesirable features that it is difficult to believe that any legislature having any knowledge or interest in the valuable fish and game resources of the state will continue it." It seems that the state institutions at that time served not public interests, but special interests. This historical appraisal gives us a picture, a baseline that we can compare with the present system.

Institutional Function for Managing Wildlife

We use three discourse-oriented techniques to examine how well the present institutional system is functioning. Political scientist John Dryzek defined discourse as "a shared way of apprehending the world. Embedded in language, it enables those who subscribe to it to interpret bits of information and put them together into coherent stories or accounts. Each discourse rests on assumptions, judgments, and contentions that provide the basic terms for analysis, debates, agreements, and disagreements, in the environmental area no less than elsewhere."[13] We begin by broadly mapping the competing discourses at play in large carnivore management and evaluating their quality. We then examine claims and counterclaims about carnivore management and their relationship to common and special interests. Finally, we use key-incident analysis to explore further the nature and quality of the discourses about carnivore management. Each technique gives us a partial test of how well the institutional system of wildlife management is working in the public interest.

Mapping the Discourse

Our first partial test evaluates the nature and quality of the discourse that is taking place. As described in previous chapters, this discourse is highly contentious. Existing arrangements determine in large part how this discourse is carried out—whether, for example, it is comprehensive and inclusive, and what outcomes and effects are produced.

Wildlife management is not a simple, linear process. It does not come with well-defined, mechanistic parts that all fit together in smooth decision making and rote practices. The difficulties often arise from differences in people's perspectives. For example, people who share a ranching background in Lincoln County, Wyoming, have more in common with each other than they have in common with wealthy second-home owners in Teton County. Each group of like-minded people shares the same physical setting, "lifestyle," economies, material goods, a relatively unique way of understanding the world, and a language for easy communication within the group. When two different groups try to interact, however, the differences in their accepted myths and narratives may be large enough that they "talk right past one another." The 90 miles between Afton and Jackson is no reflection of the vast distance between the subcultures and discourses of these two towns. In many ways Afton typifies the "Old West" and Jackson the

"New West." These differences must be appreciated if we are ever to find common ground in carnivore management. Individuals may not comprehend someone else's ideas very well because they are locked inside their own way of thinking, listening and evaluating with the assumptions of that perspective. Each group may talk about facts and symbols in very different ways, making it extremely difficult to communicate with other groups. Symbols become very important in this kind of environment.

Simon Swaffield of Lincoln University in New Zealand provided a good example of discourse analysis and the differences that exist across groups.[14] He looked at what people in New Zealand mean when they use the word "landscape." He found that the word had multiple meanings and that different interest groups used the same word in at least seven different ways. Among the management implications, he found that people use language symbolically and strategically to advocate their interests, and he concluded that the strategic use of language by different interest groups is one of their most powerful weapons against government, which largely structures and controls the management policy process. Multiple meanings make the job of government more difficult. Swaffield found that the public often uses meanings that are more inclusive, while government planners use meanings that are narrow and that exclude valid interests. This is, in part, what seems to be happening in the carnivore management process.

Table 7.1 summarizes the two main discourses about carnivore management in Wyoming and identifies the parties involved with each. In both the table and the remainder of this chapter, we use the terms defined in chapter 2 to describe the primary groups involved in large carnivore management: localists, environmentalists, and agency personnel. Recall from that chapter that although agency personnel are distinguished by their commitment to a bureaucratic and technical approach to decision making and management, in other respects they tend to align with the views of either localists or environmentalists. We discuss the bureaucratic and technical views of agency personnel later in this chapter, but table 7.1 focuses on views about maintaining or changing the current management system. The two discourses outlined in the table include dramatically different assessments of current approaches to carnivore management. Each is underlain by different assumptions, judgments, and contentions, each

Table 7.1.
A classification of large carnivore management discourses in Wyoming.

Discourse type	Kind of change needed (assumptions, judgments, contentions built in)	Nature of discourse (using language, "facts," and symbolism)
Status quo		
Held by many "localists" and state "agency personnel" (Wyoming governor's office, WGFD, Wyoming Game and Fish Commission)	"Ordinary change" is called for, i.e., maintain the status quo or entertain very small changes. The present is taken as given and generally acceptable. Change should consist of small, incremental adjustments, if any. The aim is to go slow and easy.	Maintain the present management system and remove or eliminate large carnivores if they threaten people and their established practices.
Reform		
Held by many "environmentalists" and some federal "agency personnel"	Significant change is called for. The present is taken as unacceptable. Departures from current practices are needed, including the possibility of dramatic reform.	Reform the current management system, upgrade, modernize, and find a new formula for sustainable coexistence of people and carnivores.

takes a different view about change, and each uses language, "facts," and symbols differently. The potential for misunderstanding and conflict is high in such circumstances.

Most people are locked inside one discourse looking out at other people who are using a different one. When taken to the extreme it is easy for each person to conclude that he or she is "right" and everyone else is "wrong." As Christine MacDonald noted, "Discourse is an ideological practice in the sense that it contributes to a construction of certain values and goals as more worthy than others, identifies particular institutions as primary actors in a policy issue and attributes authority to certain bodies of knowledge over others."[15] Some people, however, make the effort to understand other people's beliefs, actions, and words, and a few develop an overview of all the discourses in play and work to harmonize them and bring differing camps into an integrated and inclusive dialogue. Unfortunately, as the case studies in this volume show, large carnivore management is severely hampered by gulfs between the two dominant, competing discourses, and the current institutional system does little to address this fact or to facilitate integration and inclusiveness. Furthermore, there simply are not enough people seeking win-win outcomes in the common interest. This first test, then, indicates that the wildlife management system is not focused on the common interest per se, but is instead a complex mix of competing special interests without an effective means to integrate and include them all.

Claims and Counterclaims

Our second partial test of the institutional system of wildlife management examines the claims and counterclaims that are being made about large carnivores.[16] Discourse consists, in part, of the claims and counterclaims that people make, the values at stake, and what these claims symbolize to the claimants. Patterns in claim making over time reveal patterns in values, outlooks, and policy preferences. At present the agencies and other groups involved in carnivore management are operating without a clear picture of these patterns, which could expose sources of conflict and reveal ways to move toward common ground.

Analysis of claim making shows that there are many disconnects among people involved in carnivore management. Claimants on all sides invoke "data" of varying quality to support their views, but these data are not all equally valid. Consider, for example, the claims and counterclaims that have been made about the relationship between wolves and game animals.[17] Several groups have asserted in the press that "wolves are responsible for the mass slaughter of elk on the feed grounds, and that the wolves will eventually destroy the elk herds"[18] (the "feed grounds" are areas where large numbers of elk are fed by WGFD during the winter). One such group, Elk for Tomorrow, a sportsmen's association based in Riverton, Wyoming, and its president, Mike Rinehart, claimed in a long newspaper advertisement that the reintroduction of wolves had dangerously reduced elk calf-to-cow ratios. Based on apparently anecdotal observations, they said that information from Idaho showed that "wolves kill, not only for consumption . . . so much for the environmental claim that wolves keep the herd strong by killing off the old or defective" [*sic*].[19] This group publicly attacked Wyoming Wildlife Federation's regional representative, Lloyd Dorsey, who challenged their claims. Elk for Tomorrow said that they opposed "any plans Dorsey and the WWF have for [places] where wolves are already decimating elk herds." They claimed that "without immediate action by every caring Wyoming citizen, elk will become the endangered species, or perhaps a bittersweet memory," and they hoped the wolves would "turn their carnivorous desires towards the wolf-watching outfitters and 'eco-tourists,' the rightful target indeed." Labels of "Green Mafia" and "Eco-Taliban" were used to discredit those who supported wolves and questioned Elk for Tomorrow's claims. Another of the group's ads later that year devoted a full page to personal attacks against conservationist Meredith Taylor on the issue of elk feed grounds.[20] As a result, the debate became personalized and highly politicized.

Elk for Tomorrow has not been alone in making such claims. Friends of the Northern Yellowstone Elk Herd, Inc., of Pray, Montana, a nonprofit group that claims 2,900 members, made similar claims in a letter of February 21, 2001, to President George W. Bush. This group said that "we are not replacing the aging [Northern Yellowstone Elk herd] because of a wolf instinct called "Surplus Killing" where newborn elk calves are reflexively and wantonly destroyed by wolves shortly after birth. The other prey species are in as much, if not more peril."[21] They claimed that Mike Phillips, formerly a Yellowstone National Park wolf biologist who is now with the Turner Endangered Species Fund, had said that "the goal of wolf introduction was to drive ranchers from public lands." The group went on to say that "this is not only a violation of the Endangered Species Act, but also a violation of the 5th Amendment of the Constitution, and its 'takings clause.'" Finally, they claimed that they were "centrists and moderate in our request." Phillips denies having said what was attributed to him.[22] This discourse, like others, is based on claims of valid data, but standards of validity obviously vary. Also, observations and other data are marshaled selectively in support of claims.

Similarly, Wyoming Senator Mike Enzi claimed in a letter to Secretary of the Interior Gail Norton that wolves caused severe losses among Wyoming's elk herds, especially calves. Enzi's chief of staff Flip McConnaughey said that the senator based his letter on claims of his constituents and their reports from the Big Horn Basin and the western part of Wyoming. McConnaughey said, "People are the first ones to see it. They're the ones on the ground every day, they're the ones who watch it." These claims are frequent and widespread. They reflect the perspectives of some localists and even some WGFD personnel. The WGFD has indicated that it is also concerned about the depredations by wolves on its winter elk feed grounds.[23]

Counterclaims in the wolf/elk controversy typically come from environmentalists or federal government employees and occasionally from state employees. For example, in 2002 the U.S. Fish and Wildlife Service's (USFWS) wolf recovery coordinator, Ed Bangs, said that no scientific evidence supported the claim that elk calf populations were down dramatically in 2002, or that wolves were responsible. Some WGFD officials backed Bangs in this instance. Jay Lawson, wildlife division chief of the WGFD, said that the calving season was just over and there was no way of knowing what effect, if any, wolves had had on elk populations. He said he had received no reports that year (2002) about wolves killing elk calves. Mark Gocke of the WGFD Jackson Region said that, as of 2002, there were three packs of wolves active on the Jackson elk herd (which uses three of the feed grounds that concentrate elk in the winter) and that, even with wolf predation, the elk herd was larger than the optimum population of 11,000 animals that the state tries to maintain. In fact, the herd had reached a high of 16,236 animals in 1996, declined to 12,132 in 2001, and rose again to about 13,500 in 2002. The real effects of wolves on the elk herd are still being investigated, said Gocke. A USFWS study in Yellowstone National Park conducted since 1995–1996 shows that one wolf kills about 1.8 elk per month. Two USFWS employees monitored the number of elk killed by wolves in the Gros Ventre drainage in Wyoming for the years 2000–2002. Their estimate is that 264 elk were killed in the 12 months of 2001 out of a herd of 13,500 animals. The calf-to-cow ratio in the 13,500-animal herd was 22.8 calves to 100 cows, slightly down from the 10-year average of 25 calves per 100 cows. In addition, Mark Gocke said that research up to 2002 suggested that black bears have more impact on elk calves than wolves have. Elk are probably being affected by drought as well, he added. Overall, as Ed Bangs observed, with wolves, opinion is quickly offered up as fact. He went on to say that wolves stir more emotion than anything else.[24]

The wolf/elk controversy is but one, admittedly extreme, example of the claim-making dynamic. Most of the public either stays away from this kind of claim making or quietly sides with one side or the other. Many other claims and counterclaims are made, though, about all aspects of large carnivore management, such as whether bear baiting during legal hunting seasons near livestock increases rancher losses and endangers human safety, whether mountain lion hunting causes needless deaths of cubs, and whether the U.S. Forest Service's proposed "food storage order," designed to manage food availability in campgrounds and in backcountry areas, is really needed. The management process is immersed in a seemingly endless series of these claims and counterclaims. Some claims could be factually clarified and resolved through adequate scientific research, but many, especially those that are ideological rather than scientific in content, cannot be resolved easily.

This pattern of highly contentious claim making has harmful and long lasting consequences. It creates hard feelings among participants, increasingly rigid individual value positions, and a highly politicized policy environment. How each claim is advanced, countered, or resolved (if ever) is a reflection of the functioning of the institutional system of wildlife management. The high volume of conflicting claims that presently exists indicates that the institutional system is unable to resolve them in the common interest. It is possible that the state government and perhaps the federal government even benefit from the conflict in that managers feel free to make whatever decisions they want because it is not possible to please everyone. Agency officials in this arena have actually said publicly that they are sure they are making the right decisions because all sides are upset with them.

This second partial test of the institutional system of wildlife management indicates that the adjudication process about claims and counterclaims is not being conducted in the common interest. Instead, participants pursue their special interests. The government's and especially the state's involvement in the institutional system has not led to effective resolution of differences and may, in fact, have magnified these differences.

Incidents

Our third partial test of the adequacy of the institutional system of wildlife management examines incidents, how the system seems to spawn them endlessly and how it deals with them once they arise. *Incidents* are key events that may initially appear to be minor but actually have significant consequences.[25] They precipitate crises, focus issues, articulate people's expectations (and whether they have been violated), and clarify differences in symbolic import. Because of the political saliency of incidents (most are covered in the media), people use them to form their opinions and to provide evidence (as "facts") in their claim making. Incidents clarify how a community is divided and what it is that separates it into factions. Incidents tell us about "shared notions of what is right [which in turn] influence perception, reason, and capacity for mobilization. These inferences about what other actors think is acceptable behavior . . . are almost entirely derived from the response of key actors to the critical event."[26] Incident analysis is another window on how the institutional system of wildlife management operates. Practically, this system should help stabilize and reconcile people's expectations about how large carnivores will be managed and how decisions will be made.

The appearance of a mother mountain lion and three large cubs on Miller Butte on the National Elk Refuge just outside Jackson in winter 1999 was an incident that highlighted latent public interest in carnivores (see chapter 3). A few months later, in the midst of great public sympathy for the lions, another incident occurred: WGFD increased the hunting quota for mountain lions in the Jackson area and across the state. This decision upset many people and motivated a determined minority to watchdog the agency and criticize the new hunting regulations. These people publicly questioned WGFD's rationale for quotas, the scientific basis for its management program, and the ethical behavior of lion hunters. They also accused the agency of a lack of concern about the loss of cubs when mothers are shot, and they challenged the processes used at public meetings and in decision making. Some citizens formed The Cougar Fund, a new not-for-profit organization dedicated to seeking better mountain lion management in the West. The founders felt that they had exhausted all possibilities for redress through existing state-led institutions. The establishment of this nongovernmental organization in these circumstances is further evidence that the institutional system is not working well, at least for some citizens.

Incidents concerning grizzly bears also occur regularly. For example, grizzly bear predation in the early 1990s on Paul Walton's grazing allotment in Bridger-Teton National Forest remains a hot issue even today (see chapter 4). Another persistent incident featured prominently in the news is bears killing livestock in the Du Noir area near Dubois. Dan Engles' livestock losses, his compensation claims to the WGFD, and his criticism of grizzly bear researchers baiting and trapping bears near his cattle and cowboys without telling him is also an incident of import. Since the 1990s incidents of livestock losses from grizzly bear depredation in the Upper Green River area near Pinedale have dominated the news.

Wolves are also commonly involved in incidents. One of the most prominent was the first appearance of wolves in Jackson Hole on the National Elk Refuge in 1999 (see chapter 5). This pack was observed by thousands of people, photographed, and talked about extensively. It was featured on the front pages of newspapers locally and statewide. In addition to being interested in the wolves themselves, people were concerned about personal safety, danger to pets and livestock, and potential losses of elk. Other incidents that have flared up in the media include wolf impacts on elk on the National Elk Refuge and on state-run feed grounds. Still other incidents include single wolves killing domestic sheep and packs killing livestock and dogs near ranch houses, barns, and right in front of ranchers (e.g., in Du Noir). Wolf incidents are always contentious. In the end, many of them fade away out of public consciousness, but a few have great longevity.

One very high profile incident that occurred in Grand Teton National Park in 1996 provides particular insight into how such events play out under the existing institutional structure. Christina Cromley examined this incident, and it is worth reviewing her findings in some detail here.[27] Bear 209, a 9-year-old, 550-pound male, was captured on August 4, 1996, by WGFD in Grand Teton National Park on Elk Ranch East. A lethal injection ended his life. His capture and destruction took place in response to ranchers' claims that a bear or bears had killed 11 calves over a 3-week period in July of that year. Bear 209 had been marked previously and had a history of cattle depredation. Previous unsuccessful attempts to translocate him influenced the state in its ultimate decision to kill this animal, even though it was a federally listed threatened species and was trapped in a national park. The National Park Service supported WGFD's decision to kill the bear, and the park responded to the public outcry with a management summary of its justification for killing the bear.

As Cromley explained, the killing of grizzly bear 209 became an incident because it symbolized many problems in the larger management policy process. It permitted battle lines to be clearly drawn. Widespread public reaction followed the killing of this bear. Newspapers printed articles, editorials, and letters. Correspondence and personal communications took place between the superintendent of Grand Teton National

Park, conservation groups, and unaffiliated citizens. Letters were written to the Secretary of the Interior and the head of the National Park Service. A citizens' group organized a petition protesting the killing and collected 831 signatures from residents and tourists in just a few days. Environmental groups mentioned bear 209 in their newsletters. An epitaph for the bear appeared on the World Wide Web. The incident polarized, alienated, and mobilized a great many people, resulting overall in bad press for the agencies and a drawdown of public trust.

The incident assumed larger-than-life proportions and accumulated more and more symbolic significance. The death of bear 209 was invoked by environmentalists to question whether livestock grazing should continue in Grand Teton National Park on leases that had expired under the original authorizing legislation but were extended by the park superintendent and a congressional bill. Others used the incident to question the professionalism, leadership, and decision-making arrangements of the WGFD, the National Park Service, and the USFWS. They questioned whether killing bear 209 was in the common interest or was done for special interest reasons. They claimed that agency actions favored a very prominent rancher in the valley at the expense of grizzly bear conservation.

Cromley went on in her analysis to explore the context of this event. Examining the participants and perspectives revealed through the incident, she found that people's expectations differed dramatically about (1) when it is appropriate to move bears, (2) the management zoning system, (3) allowing grazing in a national park and the relationship of grazing to wildlife, and (4) interagency decision making and participation in decision making. She explained these differences as a clash of myths that left many people feeling deprived in certain ways—disrespected, powerless, betrayed. This is why there was such a dramatic public response and vehement debate. She noted that countless decision processes intersected in the final choice to kill bear 209, and her identification of these multiple processes affirms how complex the institutional system of wildlife management actually is. Her recommendations were to clarify expectations and demands about actual bear management procedures and about interagency coordination efforts and participation in decision making.

This last partial test of the institutional system's operations indicates that incidents feature prominently in management and have negatively affected the public's view of the legitimacy of agency operations. The large carnivore management process continues to lurch from incident to incident, with inter-incident periods filled with disconnected discourse and opposing claims and counterclaims, against a backdrop of traditional management routines.

Taken together, the structure and functioning of the institutional system of wildlife management, as evidenced through laws, regulations, meetings, and agency decisions, as well as through discourse, claims and counterclaims, and incidents, show a conflict-laden and highly politicized process with few prospects for improvement. These three partial tests of the adequacy of the institutional system illustrate that it operates suboptimally. This kind of situation, according to policy scientist Ron Brunner of the University of Colorado, constitutes a crisis. "Crisis is not the result of conflict, but of the failure of means for resolving conflict." In short, the suboptimal performance of the present institutional system of wildlife management "is not the result of special interests dividing the community, but rather of particular maladjustments which prevent compromise between these interests," in the words of social scientist Carl Friedrich, who wrote on this kind of problem decades ago.[28] Understanding why this is so is essential if we are to find ways to upgrade performance.

Explaining the Institutional System of Wildlife Management

The institutional system of wildlife management can be explained by understanding the perspectives and practices of the people and organizations involved. These are "abstracted," so to speak, in the narratives (stories, beliefs) at play that comprise the basic formula for action, in the bureaucratic, managerial orthodoxy of the agencies, and in the assumptions of professionals and the roles they play. With this view of how and why the institutional system behaves as it does in large carnivore management, we can begin to develop and implement strategies to improve performance.

Beliefs and Narratives

At the heart of all cultures and human behavior are narratives about people's basic beliefs. Everyone communicates through stories of one kind or another. These narratives or stories tell people about who they are, what is important, and why they do what they do.[29] Narratives abstract and mirror the doctrine (basic beliefs), the formula (rules or code for behavior), and the symbols of identification in a person's or culture's perspective (see chapter 2). Institutions directly manifest these narratives and basic beliefs. To understand the institutional system of wildlife management, we must ask ourselves what the core story is, who has the most to say about it, and whose interests or values are most served by the story. We must also ask what competing narratives, if any, might exist and how these might affect the institutional and policy dynamic of wildlife management.

Core Beliefs

Core beliefs can be determined by examining the record of an individual or organization over long periods and looking for recurring themes and central tendencies in actions and words. Because of its dominant position in large carnivore management in the state, WGFD deserves close scrutiny. The core belief within the department that stands out, based on our observations over three decades and the work of other analysts (see chapter 2), is that power and the competition for power are of paramount importance. This belief provides a formula for how the department hoards or shares authority and control and with whom. It dictates that WGFD should have authority and control over all wildlife within the state, a classic states' rights assertion. This core belief figures into the department's identity, its expectations about the world, and the demands it makes. Large carnivore issues reveal this core belief in the department's interactions with other people and organizations and in its internal operations.

One event that gives a clear window into WGFD's dominant narrative and core beliefs is the suspension of department biologist Dave Moody in April 2003 for comments he made about a new Wyoming state law dealing with wolf management.[30] Moody observed at a professional meeting (the 15th annual North American Interagency Wolf Conference) that the new law is "extremely problematic" in that it "does not provide long-term, adequate protection [for wolves]." According to newspaper reports, Moody was suspended from his duties because of these comments, which directly challenged the state's beliefs, narrative, and actions. In many respects Moody showed the state's position for what it was—an attempt to assert power and make a claim for increased authority and control over wolf management.

Harry Harju, recently retired assistant chief of WGFD's wildlife division, said in a perspective piece in the *Casper Star-Tribune,* the state's most widely read newspaper, that "one way to shut agency employees up is to make the director a political appointee. . . . This discourages knowledgeable professionals."[31] Directors will censor "actions and words of department employees for fear something an employee does or says will cause removal of the director." Commenting on the Wyoming Game and Fish Commission, Harju said that some political appointees have "little interest in wildlife, and were mainly representing special interests that promoted them as commissioners." Furthermore, these commissioners "were mostly interested in complaining, nitpicking, threatening and harassing employees . . . and voting against things that were important to wildlife but disliked by agriculture or industry." Such a style of leadership and climate for wildlife management could account for incidents such as Moody's suspension.

Although WGFD subsequently said that Moody had only been "placed on paid administrative leave for a short time" and "was never suspended," this is not the first time that WGFD officials have attempted to control their staff professionals in support of the state's beliefs (for example, see the parallel story about Joe Bohne, WGFD biologist, in chapter 3). Such incidents show a pattern of state officials enforcing compliance and adherence to their states' rights ideology. The value at the center of these incidents is power—power over employees and control over their knowledge and its dissemination.

More broadly, the dominant sector of Wyoming's culture has put the state at a crossroads, according to Paul Krza in a *High Country News* article (see chapter 2).[32] Wyoming has traditionally been very conservative, adhering to an inflexible states' rights ideology. It has catered to local agricultural, mineral, and other special interests in its decision making. Wyoming has consistently sought to maintain the status quo and to maintain or increase its power dominance in the wildlife management system. Progressive wildlife management, open discussion, inclusiveness, and integration have often been casualties of this agenda. According to Krza's article, for example, former governor Jim Geringer (1995–2003) "squelched any dissent from within the state agencies" against his antienvironmental stance.[33] Agencies such as WGFD were not allowed to operate independently, even if they did have environmental concerns. The former state auditor, Dave Ferrari, said Geringer ran his administration like "an arrogant know-it-all kind of guy who was argumentative and didn't want to hear anybody else's point of view."[34] Present governor Dave Freudenthal has said that he wants to change things, especially agency behavior, but has proceeded very slowly and has appointed agency leaders who have a history of resistance to change.[35]

The Importance of Narrative

According to psychologist Jerome Bruner, the narrative way of knowing (that is, attending to the story line) is one of two basic ways in which people comprehend the world; the other way is "logico-scientific."[36] Most of the ongoing discourse about large carnivore management uses the narrative way of knowing; only a very small part of it is about the logico-scientific basis of wildlife conservation. Dave Moody's comments about the state's wolf management plan were based on logico-scientific knowledge, for instance, but they came into direct conflict with the state's narrative way of knowing, which emphasized power and control.

To identify the dominant narrative and counter narrative in policy debates about carnivore management, we looked for recurring subjects or themes in these debates and observed how this was reflected in language used and in actions. The identification of key themes was based on their repetitious appearance in the data. Our characterization of narratives and themes is admittedly an oversimplification of a very complex dynamic, but it provides a plausible explanation of why the institutional system of wildlife management currently operates in the ways that it does.

We identified two main stories competing for dominance over policy making, one told by localists and state agency personnel, especially WGFD, and the other told by environmentalists and federal agency personnel (table 7.2). Historically, localists (ranchers, hunters, and others who share the beliefs of the Old West) and state agency personnel (especially WGFD and agricultural interests) have been allied and have controlled wildlife management in the state. Not surprisingly, their values and their story dominate, and they are the ones most served by the institutional system today. In the last few decades, however, a counter narrative promoted by environmentalists of the New West and some federal agency personnel has gained some prominence, especially at the national level, and is being implemented on some federal lands.

Table 7.2.
Key themes in the narratives about large carnivore management in Wyoming.

Dominant narrative: Localists and many state agency personnel	Counter narrative: Environmentalists and many federal agency personnel
Large carnivores are not wanted by locals or the State of Wyoming.	Large carnivores are wanted by the broader American public, and restoring nature is the "right" thing to do. Conservation science dictates that carnivores must be protected and provided with more habitat.
The highly individualistic, locally controlled, competitive environment is the "Wild West" and that is what we want.	All Americans are part of a larger family and must work for the overall best interests of the nation.
Large carnivores create unacceptable problems for ranchers, hunters, and human safety.	Any problems can be managed. Overall, large carnivores pose no great risk to people, livestock, or wildlife.
Local ranchers and rural residents are the ones who have to live with large carnivores and suffer harmful impacts. We should have the final word on the matter.	The impacts suffered by locals are not severe, and the costs can be minimized. Localists are provincial, poorly educated, narrow-minded, and self-interested. "Objective science" should have the final word on the matter.
Carnivores are just another example in a seemingly endless series of events in which outside people force change on locals against our will.	Carnivore conservation is the will of the American public, and locals can be fairly represented and compensated if they want to work with us. Localists need guidance and modernization, with the aid of science and enlightened management.
We are the "David" in a "David and Goliath battle" (the federal government and its environmentalist allies are the "Goliath").	"Goliath" in this case is the good guy and represents the American people. It may be possible to balance local interests with national interests, but if not, national interests should prevail.
Outside people have declared "war" on us and we will fight to the end.	In the end, national interests and conservation science *will* prevail. Localists and state officials do not want to cooperate and work out problems together rationally and realistically.

This counter narrative calls for a different institutional formula, and it promulgates less dominionistic, utilitarian beliefs about our relationship to nature and a different view of the relationship between policy and science. It also distrusts localism and states' rights as the best formula for public management of natural resources.

Dominant Narrative

The dominant narrative says that power and influence must be exerted to maintain control of institutions and decision making so that local and state interests will be served. It also says that federal intervention is not only undesirable, but it is unacceptable and must be resisted in all cases. This view envisions environmentalists and federal agencies as all-powerful outsiders who are set on subduing and ultimately destroying local and state interests.

Adherents to this narrative tend to believe that outsiders, especially the federal government but also its environmentalist allies, have an evil master plan to overwhelm localists and the state government. The "feds" and environmentalists are out of control, spreading like weeds (or wolves in this case), creeping and destroying their way throughout the West's culture and resources. Many ranchers, hunters, state game and fish personnel, county commissioners, and local business owners seem to believe this story, as do many policy advisors and decision makers at the state and local levels. Even some members of the media, who may perceive of themselves as objective and neutral, believe it, as do some federal employees who work to further states' rights. The notion that the federal government and environmentalists are evil has become an unquestioned, albeit tacit, assumption that influences most decision making and discussion of alternative ways of managing resources. Localists even liken federal government initiatives to communism, implying a move from freedom, economic or otherwise, to oppression. The solution that this group would like to see is for the federal government simply to go away, giving all management authority and control to the state and the localists.

The effect of this rhetoric about "local control" is to mask the self interest of these participants. Built around a metaphor of local rights and the need for a fair playing field, this imagery is used to persuade policy makers and public opinion that the current situation is unfair and needs to be fixed according to the localist formula. This story says that the playing field is tilted in favor of outside interests from the east and west coasts (both perceived as environmentalist strongholds), thus preventing local interests from competing fairly. Localists say they are caught up in a game that they have no chance of winning, through no fault of their own. Since public sympathies often lie with the underdog, they portray themselves as David in a fight with Goliath (the feds and environmentalists).

According to this narrative, federal government intervention is always bad. Reintroduction and protection of large carnivores is a game with unfair rules, which are not always clear. The feds and environmentalists are ramming something down local throats that people do not want. These opponents are all-powerful, in effect acting as police, judges, and jury all in one, with no chance of appeal. This group feels that it can be victorious only by fighting tooth and nail to the end.

Counter Narrative

Whereas the dominant story draws on the metaphors of individualism, freedom, business, and local control, the counter narrative has the key theme of rational policy making driven by sound science. The story line here is that large carnivores are desirable and that the federal government should intervene to ensure that they thrive. Adherents have attempted to convince policy makers of this by characterizing policy making as scientific, rather than ideological. They say that the true test of any carnivore policy proposal is its factual biological basis. By focusing on the biological facts, promoters imagine that they are presenting objective truths rather than engaging in subjective, political debate. The appeal to objectivity reduces policy making from a complex, value-laden process to a simple matter of applying an objective formula (in this case, the positivist science of expert biologists). To provide support for their claims of objectivity, those promoting this story attempt to connect their preferred policy outcomes to the national public interest. For example, they may conceal their own self interest behind arguments that implementation of the Endangered Species Act is logical and serves the broader public interest.

Adherents of this counter narrative believe that once carnivores are well established there will be a reasonable chance of achieving larger goals such as ecosystem management and restoring nature itself. This story is full of the symbolism of modern science, including pro-rationality terms such as "restoration ecology," "ecosystem management," "endangered species conservation," and "ecological connectivity." The story is also heavily laden with morality-based rhetoric, such as the "land ethic," using Aldo Leopold and other respected figures to justify a policy preference for restoring nature, which is considered to be inherently good—the right thing to do (a redemption narrative). As in the dominant narrative, particular symbolism and vocabulary are used to tell a certain story that proponents want policy makers to accept as the basis for decision making. In this story, environmentalists and federal management agencies consider themselves rational, considered, deliberate, logical, and public interest-minded. They see the localists and state agencies—who view themselves as heroes taking on the risks of life on the frontier in the wild West—as provincial, poorly educated, narrow, and self-interested people who are in need of guidance and modernization, with the aid of science and enlightened management.

According to this counter narrative, federal government intervention is not only desirable, it is imperative. Carnivore restoration is a good thing, but it must be implemented in a way that ensures that the national interest is achieved, even if this means overriding local interests. The guiding principle of scientific objectivity will lead the nation to the optimal resolution, and it should be left to science to determine the correct solution. To do otherwise would be to sacrifice the national interest to the local, narrow self-interests of ranchers, hunters, and the state. Issues of morality and justice raised by localists should not be used as the basis for decision making.

Conflicting Narratives

When the dominant and counter narratives in carnivore management are examined together, it is apparent that the central motif of both is the issue of power (authority and control). The foundation of the conflict is the appropriate balance of local and national interests, a long-standing clash that goes back to the founding of the republic.[37] At issue is who should have power, how it should be used, and importantly, whose interests or values it should serve. Thus, the conflict about large carnivores is yet another manifestation of an ancient contest over whose values and narratives will guide decision making. This conflict, intense and basic, has profound implications for each side, for how the institutional system will operate, and for how we will govern ourselves generally about environmental matters in the future.

At present, the institutional system of wildlife management functions under the dominant narrative in favor of certain core beliefs. It privileges WGFD, its states' rights ideology, its past practices, and the beliefs and interests of localists. Although the symbolism contained within the competing story is vivid and the plot compelling, it has generally failed to alter how policy makers (at local and state levels) have constructed their own stories and act on them.

Bureaucratic, Managerial Orthodoxy

Carnivore management is largely defined and shaped within bureaucratic government agencies, which set tight boundaries on how people are allowed to interact in the management arena.[38] For example, in some public meetings held by agencies ostensibly to gather public input, speakers have been limited to 2 minutes to make a statement. At one meeting on mountain lions sponsored by WGFD in June 2001, public comments to the assembled group as a whole were forbidden—the only permitted forms of "input" were comments written on flip charts or private conversations with state officials (see chapter 3). These formats restrict public discourse and leave many participants feeling insulted and powerless in what they believe should be an open, deliberative, democratic process.

Bureaucratic organizations are built on certain assumptions about appropriate power relationships among government and citizens, professionals and laypeople, and the official and unofficial parts of policy making. The organizational literature shows that bureaucracies usually operate in self-serving, undemocratic ways, to maintain power in the hands of the few rather than the many.[39] Organizational researcher Charles Perrow argued that the biggest danger in such an organization is "how it inevitably concentrates those forces [social resources] in the hands of a few who are prone to use them for ends we do not approve of, for ends we are generally not aware of, and more frightening still, for ends we are led to accept because we are not in a position to conceive alternative ones."[40] Bureaucracies, when left to themselves, typically deem only a limited set of alternatives feasible to consider in addressing problems, and these typically support and maintain existing bureaucratic structures and modes of operation. These limitations affect how participants interrelate, how decisions are made, how resources are allocated, and how work is performed. They trap people, including bureaucrats themselves, in complex management structures that cause, magnify, and recycle destructive conflict.

Bureaucratic wildlife management agencies often exclude people and fail to integrate information to find common ground. The reasons for this are well documented. First, bureaucracies have built-in, "default" tendencies, and during times of stress they tend to revert to these to guide their behavior. When confronted by public demands to be more open or to consider other data, wildlife agencies may do just the opposite, retreating to their core defenses and standard operating procedures.[41] Bureaucrats calculate what they should do or how they should behave in any given situation based on their past experiences and these built-in tendencies. They make the best guess about what will appease powerful interests who are aligned with them or who control their fates, and then by persuasion or other means they try to enlist sufficient support to have their decisions "legitimized."

Second, bureaucratic agencies work very hard to maintain their "turf," their independence, and their decision-making authority. Power relations among resource management agencies have always been very important, as have the relations between these agencies and other powerful groups—congressional delegations, county and municipal governments, environmental groups, and industry. The preoccupation of agencies such as WGFD with authority and control is evident in their press releases and behavior with other actors (as documented in previous chapters).

Third, the complexity of the institutional system for wildlife management has increased dramatically, which impedes effective management by bureaucracies. The institutions, organizations, policies, and practices of numerous other arenas intersect and influence wildlife management, among them economic, agricultural, and recreational policy. Overlapping and fragmented jurisdictions and limited resources contribute to the complexity; missions, controls, and boundaries are confused. Bureaucratic agencies are not adaptive enough, and their staffs do not have the requisite skills and knowledge to function effectively under these circumstances.

Fourth, the context in which the agencies must operate has also become far more complex. The public is growing and diversifying, and its demands are intensifying. Structures of governance are also growing and diversifying. The effect is to multiply divisions in our communities and heighten competition among special interests, stemming, ultimately, from profound ideological differences that will not go away. These differences play themselves out through issues such as large carnivore management. Symbols become especially important in such situations. Carnivores have become symbols of "big bad government," "untamed wilderness," and much more, depending on who uses them and for what purposes. Again, the bureaucratic mode of organization is ill suited to deal with such complexity.

Because of these and other bureaucratic behaviors and limitations, it is difficult to build an integrated and inclusive regional institutional system to address wildlife problems. The wildlife bureaucracy divides rather than unites the community. Bureaucratic practices lead to conflict, but it is not conflict itself, but the failure to resolve conflict, that is the problem. This problem is talked about as "gridlock," "log jams," or "train wrecks." People's frustration stems from the inability of agency bureaucracies to integrate and include the public effectively in ways that actively facilitate the finding of common ground. The institutional system—that is, the people, organizational arrangements, and formal and informal rules at play—in too many cases does not permit thorough, timely input, adequate public involvement, or the upgrading of knowledge from relevant disciplines to be included in deliberations, planning, and management decisions.

What is the solution? Expanding confused bureaucracies is not the answer, although this is what we often do. Government alone cannot mediate the differences among interests, in part because of its own weak leadership, adherence to rigid policy preferences, and the tendency to over-control agendas. A consequence of all this is that special interests (which sometimes include the agencies themselves) are able to limit each other's actions through bureaucratic "red tape," litigation, media attacks, land use planning forums, business practices, influence peddling, or other means. In such a climate it becomes difficult to find agency and elected officials who are willing to stand up and be responsible and accountable for their decisions. This again leads to a loss of trust and faith in government. To improve wildlife conservation, especially large carnivore management, bureaucracies must be reformed.

Professional Roles

Professionals staff the agencies that dominate wildlife management at both state and federal levels. They tend to operate under certain shared assumptions, which strongly influence their behavior as they develop and implement carnivore management policy.[42] The assumptions and resulting behavior are often problematic for efforts to find common ground.

The "professional" approach to wildlife management today is strongly grounded in two notions. The first is that biological data are sufficient to supply the information needed to manage wildlife—or, in the words of Dan Decker, president of The Wildlife Society, and his colleagues at Cornell University, the insights from the biological sciences are the "nearly exclusive keys to best management decisions." This assumption dictates that biologists should be given a special place, a privileged central role, in decision making. It also means that public demands as well as other sciences, particularly social sciences, are largely irrelevant to the work of these experts and to wildlife management generally. The second, related assumption is that government biologists and managers, because of their training, experience, and positions, are the only people with the qualifications and the authority to make appropriate wildlife management decisions. In the extreme version of this notion, these professionals do not need the "help" of the public or scientists from other disciplines.

Both of these notions, which are seldom acknowledged even at professional meetings or by professional societies, have significant consequences for how wildlife is managed and who gets a say in management. Both assumptions lie deeply buried in the basic philosophy that underlies the wildlife profession, game and fish management organizations, and the overall institutional system of wildlife management. They are often "invisible" to the professionals themselves and to others who work with them.

The tension between professional experts and democratic governance is an important political dimension of our time and clearly at play in large carnivore management. Democracy and its emphasis on equality of citizenship, public input, and freedom of choice sometimes clashes with officials who may "over control" decision processes. On the one hand, agency biologists understandably want to use their high professional standards to make decisions about public resources, drawing on their expert knowledge and judgment. But social scientists have found that technical experts in bureaucratic positions actually tend to circumvent the democratic process, while justifying their actions to themselves in self-serving ways. They and their agencies may claim that they are apolitical, that their calculations are accurate and objective, but the record shows that agency experts often fail to include democratic considerations as well as other important features of the context. This has led to increasing public antipathy toward bureaucracies. Citizens are entitled to have a say in decisions about public resource use. On the other hand, it is also clear that citizens do not always have the knowledge and skills to participate meaningfully in complex decision making about public land and resource management. They may not be knowledgeable about the technical matters involved or the broader implications of their choices, and sometimes the reason citizens want to gain influence is to secure decisions favorable to their special interests. This can lead to shortsighted, parochial decision making. Public participation is not a "silver bullet" that will solve all carnivore management problems. Nevertheless, ordinary citizens are capable of a great deal more participation than is generally recognized or acknowledged by agency officials and professionals.

In summary, what we have today is an institutional system that is not working very well in the public interest. It is less than successful in managing mountain lions, grizzly bears, and wolves. States' rights and local beliefs, evident in narratives and deeds, bureaucratic arrangements, and privileged professional expert roles, dominate the wildlife management system. Community relationships are emotional and full of misunderstandings, management decisions take place without relevant data and assessment, using procedures that are unfair and that lead to mistrust, and special interests vie for influence based on fundamental beliefs about what is right and wrong or how the world should be. All this leads to highly adversarial interactions and winner-take-all tactics. Information is used as power, there is little direct contact among interest groups, and beliefs are firmly set as matters of principle. This in turn leads to personal antagonism and defensiveness, inflation of issues (e.g., carnivore management becomes an issue of property rights and individual freedom), breakdowns in communication, spiraling mistrust, and polarization into opposing camps. The current structure and functioning of the institutional system favor continued state and local dominance. There is little incentive for those in control of the system to change, but the costs of this arrangement to wildlife and to the common interest are high.

To improve wildlife management, the traditional relationship between the professional expert and the public must be adjusted.[43] The key for wildlife management in the 21st century, according to many people, is an adaptive management approach that builds on the strengths of conventional management, yet moves well beyond the assumptions and present tension between wildlife professionals and the public. What is needed are new precepts, a "civic-minded" kind of professionalism that seeks integration and inclusion and has the knowledge and skills to facilitate this end.

Fortunately, a few professionals are beginning to give up their traditional assumptions in favor of this more pragmatic, effective professional style. A new paradigm for professionalism is developing rapidly in some regions. It is much more integrative, participatory, practical, and grounded in both experience and theory.[44] Its principles focus on effective integration of information from multiple disciplines (e.g., wildlife management, economics, politics, sociology, and integrative sciences) and the inclusion of diverse stakeholders in more deliberative decision making. It seeks cooperation, understanding, trust, shared information, a conciliatory approach, frequent contact among all interested people, and development of mutual respect. This new paradigm, discussed and tested in a few situations in the West and elsewhere, offers ways to deal with many of the problems so clearly evident in the present institutional system of wildlife management.

Recommendations

We recommend three strategies that we feel are both promising and pragmatic ways to achieve the overriding goal of coexisting sustainably with large carnivores. The first is to make "practice-based improvements," or smaller scale interventions in the field, working cooperatively with ranchers and others to solve specific problems, such as carnivore predation on livestock. Opportunities abound for this strategy. This approach can help to minimize real problems and encourage everyone to see carnivores (and each other) in a new light. Second is to upgrade leadership, at all levels and in all forms, to move institutions actively toward more effective performance. Developing "transformational leadership" styles will require leaders to learn explicitly and systematically about institutional systems and more, but this strategy can help reform institutions to be more integrative and inclusive. Third is to initiate changes in the dominant policy narrative, the basic beliefs and story of the existing institutional system and regional cultures. A new "meta-narrative" could include more people, a broader range of beliefs, and more widely shared interests; its core message should be health—in the broadest sense, for individuals, communities, and landscapes—and responsibility, both collective and personal.

Practice-Based Improvements

Large carnivore management policy consists of a complex system of interconnected, smaller decisions. It would be extremely difficult, if not impossible, to understand how all the relevant decisions actually function to affect large carnivores. Accordingly, the key to improving wildlife management is to focus our limited attention and resources on select components of the management system.[45] The "practice-based" approach uses actual experience rather than theoretical principles as the basis for making improvements. Much of the current debate over carnivore policy assumes that government can improve policy from the "top down." This is partly true; after all, it seems obvious that large carnivore management is a large-scale problem. Although a top-down approach such as a national or statewide management policy would be helpful, the improvements must ultimately occur at the "bottom" or operational level. Indeed, many more opportunities for change (and for learning from change) exist at lower levels, testing them involves lower risks and lower costs, and implementing them is quicker and less complicated than making systemwide changes. The participatory projects described in detail in chapter 6 provide some examples of a practice-based approach.

There are three components to a practice-based strategy to reform the institutional system that manages large carnivores: (1) find and describe successful management practices and programs, (2) adapt and diffuse them widely, and (3) open up new opportunities to build additional program successes. Once individual and local programs that appear to be successful have been identified and described, they need to be reviewed to see how they have aided large carnivore conservation. This critical appraisal will explain the formal and effective reasons for a program's success. This strategy shifts attention to a constructive, positive focus that can motivate and inform actions for large carnivore conservation on an ongoing basis. For example, a developer might take large carnivores into account when planning a new subdivision, or a county government might devise an open space policy for ranch lands that also supports large carnivore habitat and minimizes predation on livestock. Successful field-tested models can be used to set best practice standards and can be adapted and replicated in other locales. One good example is the educational work of Patricia Sowka, a Montana wildlife biologist, and Jamie Jonkel, of Montana Fish, Wildlife and Parks, who have developed "Living with Predators Resource Guides."[46] These guides are designed to equip people who live, work, and play in grizzly-occupied areas with the information and tools they need to coexist successfully with bears. The outreach opportunities related to this project are huge. Outfitters' associations, outdoor groups, real estate businesses, homeowner associations, and others could use these educational materials.

Our three-part strategy could be implemented immediately. Several successful management cases merit study, including the work of Mike Jimenez (USFWS) on wolves, Mark Bruscino (WGFD), Barb Franklin (U.S. Forest Service), and Steve Primm (Northern Rockies Conservation Cooperative) on grizzly bears, and Timm Kaminski (Mountain Livestock Cooperative) on wolves and bears where private lands meet public resources.[47]

A 2001 guest article in the *Jackson Hole News* by Steve Primm and Louise Lasley said that the practice-based approach "turns conventional planning on its head. Instead of starting with a colossus of words that pretends to dictate what happens with grizzlies, people and land, it starts with small-scale action to solve specific problems. What we learn from these actions can then go into a written plan. The results: A plan based on reliable knowledge, and a plan that people trust because they know the stuff that's in it really works."[48] In another article, Primm said that with this "intelligent process, there is no need to attempt to design a comprehensive 'blueprint' for how and where to manage grizzlies. Instead, plans can be developed sequentially through a series of adaptive experiments in grizzly conservation."[49] Primm and a growing number of other people see that this approach is the key to long-term coexistence with large carnivores (see chapter 6). This is also the approach being used by Seth Wilson of the Northern Rockies Conservation Cooperative in central Montana as he works closely with ranchers and others to develop sound ranch management practices in grizzly country.[50]

Each of these field efforts, educational projects, or other innovations can be considered to be a "prototype." Although prototypes are generally set up to generate knowledge on which to base future improvements, many of these cases were not established specifically for that purpose. Nevertheless, by thoroughly appraising such cases, we can learn an enormous amount about what works and what doesn't work. Drawing on these lessons, participants in other locations can invent, evaluate, select, and implement alternatives for better conservation and management.

The second part of the practice-based strategy is to disseminate the lessons from successful prototypes as widely as possible. Case studies or stories can be one source of information made available to people in other areas, who can then modify the lessons to suit local needs. Best practices that are cooperative, locally based, inclusive, and participatory offer the greatest promise of success. State and federal governments can provide leadership by endorsing and supporting prototyping. In large carnivore management, a field that has few ongoing, systemwide mechanisms for appraisal, prototyping and dissemination of best practices can be especially helpful. Workshops, demonstrations, site visits, educational programs, and other means can all be used to disseminate best practice standards.

The third part of the practice-based strategy is to facilitate new opportunities for better carnivore management politically, geographically, and institutionally. These could include not only informal groups that come together to tackle specific local issues, but also innovative ways to engage new people and capture creativity, energy, and the desire to get things done. This requires effective leadership. Resources could come from stopping some present activities that have only limited or short-term benefits to large carnivores and people associated with them. Instead of pouring money into additional detailed habitat studies in some local situation or new geographic information system maps that may or may not help decision makers, put resources into creating new opportunities to solve problems on the ground in cooperative ways. Additionally, the idea of combining economic strategies for private and business interests with large carnivore conservation must be considered seriously and used beyond compensation and damage payments. For example, a workshop in Bozeman, Montana, in December 2003 queried ranchers, environmentalists, and government about opportunities to address the problems of large carnivore management, including economic incentives for changes in ranch operations.[51] The workshop identified four areas of potential agreement: (1) the use of small-scale conservation projects to try out new methods involving locals to demonstrate success; (2) the creation of incentives to modify ranch operations to accommodate carnivores; (3) the need to get more managers into the field with locals to prevent or ameliorate conflicts with carnivores quickly; and (4) the need for peer review of science and its management interpretations by critical independent bioscientists.

To implement a practice-based strategy, two things must happen. First, to increase the likelihood of success, a process for appraising carnivore management programs and practices should be organized. A useful appraisal process will depend on establishing creative means to harvest lessons from past experience. There are numerous successful, progressive, problem-solving exercises underway in the West today that can suggest valid criteria (in terms of concepts and practices) for appraising programs. As well there are many knowledgeable practitioners and researchers in both the social and biological sciences, as well as the integrative, holistic sciences, whose combined expertise could be channeled into appraisal. For example, traditional public and private funding patterns that favor biological research and short-term management solutions can be redirected toward practice-based approaches, such as prototyping.

The second thing that must happen is better communication among state and local efforts and all other parties. We clearly need an effective means to diffuse the vast amount of professional and organizational experience and to overcome policy, science, and management hurdles. For example, reducing the intergovernment (i.e., state and federal levels) power contest over management authority and control of policy and programs would help immensely. Diplomatic, educational, and financial strategies to do just that are available, but are not currently being used.

Enhanced Leadership

Effective leaders are urgently needed in wildlife management, not only at the top of government agencies, but at all levels, in local associations and environmental groups, in the field and in the office. Leaders can change institutions for the better if they are motivated, skilled, and committed.[52] The challenge for modern leaders is to find paths of common interest—or what might be called cooperative, problem-solving processes—that are environmentally and socially sound. Leaders must be capable, knowledgeable of environmental and development issues, and sensitive to both the public and private sectors.

Specifically, leaders should have the ability to carry out more comprehensive, contextual, and rational wildlife management programs than currently exist. They must understand how to design and implement practice-based strategies. They must be skilled in integration and inclusion. The activities and problem-solving approaches of good leaders are well known, well tested, and widely described, but still are little used in carnivore management. Too often we recycle the same old unworkable methods or approaches—conventional thinking, bureaucratic arrangements, and ordinary problem-solving approaches—while real problems continue to grow and press on us. We must overcome these ineffective ways of addressing problems.[53] There are, of course, many social factors in addition to leadership that play a critical role in the success or failure of wildlife conservation. Nevertheless, having leaders who are critical thinkers, holistic observers, skilled managers of people, and users of a host of technical tools, all designed to aid in finding common ground, is essential.

A New "Meta-Narrative"

We need a new, more integrative and inclusive meta-narrative to achieve a higher level of institutional effectiveness. A meta-narrative is a narrative about narratives, a story that encompasses and explains other, smaller, localized stories. It is a story that people can use to recast policy problems. It offers a new conception to which embattled players might subscribe and opens up new possibilities to solve formerly intractable problems. The practice-based strategies and enhanced leadership skills discussed above can help to bring about such a meta-narrative. The power of a meta-narrative is perhaps most apparent in times of crisis, such as when threats to national security are clear to virtually everyone. The "national security" meta-narrative then overrides all other local, special interest narratives, even formerly competitive ones, and helps unify people's thinking. Diverse people who ordinarily might hold different narratives are willing to subordinate their special interests to the overall common interest. The new meta-narrative for large carnivore management, we hope, will focus our attention and efforts on healthy human and carnivore populations, community integrity and sustainability, and individual and collective responsibility.

As we have seen, narrative analysis of large carnivore management shows how particular stories or policy narratives dominate the conflict-laden management process we see today. All of the current narratives are valid, given the origins of the communities that spawn and perpetuate them, but none goes far enough to encourage an integrated, win-win outcome for the larger community. A new meta-narrative could be directly helpful in adapting the institutional system and its complex policy dynamics toward common interests. One way to accomplish this is for all parties—localists, state and federal agency personnel, and environmentalists—jointly to develop an integrative meta-narrative though extensive practice-based engagement. That is, by working together on a series of smaller, on-the-ground projects to solve specific problems, participants with different perspectives will, over time, begin to forge a new, shared understanding (a meta-narrative). Like recruits going through boot camp, they will eventually come to see themselves as an effective team working toward a common purpose. They will build a new path that they will travel together. Policy meta-narratives have a strong prefigurative effect, conditioning the thinking of all involved. It is only through such an approach that the basic underlying policy narrative can be adapted.

Policy analysts who have investigated the power of meta-narratives have concluded that the way to change institutions for the better is not to develop a critique of existing policy. The logico-scientific approach is no match for a powerful policy narrative. The best way to bring about change is to undermine a policy narrative by creating an overriding counter narrative. According to professor Emery Roe, the meta-narrative serves to distance its listeners from their original stories. It acts as a departure device, and in some ways it is the antidote to conflicting, poorly performing narratives and institutions.[54] Thus far, no meta-narrative has emerged or been advanced in large carnivore management that is sufficiently compelling—in explaining the conflict and at the same time suggesting integrated, win-win solutions that can move us forward—to engage most if not all parties. We feel that the best way to begin the process of finding a new, integrative meta-narrative is through practice-based engagement. This approach can demonstrate a narrative of cooperation, mutual respect, shared goals, and coexistence through actual progress on the ground. Such a meta-narrative might encompass recognition that further conflict is counterproductive and undesirable. Its success will depend in large part on having in place a large stakeholder group that is willing to embrace it without much resistance. The key is to find a story that almost everyone can believe and support.

Conclusion

The institutional system of wildlife management used to restore and conserve large carnivores is weak. Its performance is suboptimal, and its functioning seems to be deteriorating as the context becomes more complex. Data from discourse analysis, patterns of claims and counterclaims, and incident analysis amply demonstrate this fact, as do the management reviews in preceding chapters. The institutional system's present structure and mode of operation trap many people in patterns of destructive conflict. It is clearly focused on power and most often serves the state, WGFD, and local special interests at the expense of broadly shared or common interests. There are many reasons for the persistence of this weak institutional system, including cultural beliefs (narratives), bureaucracy, and rigid expert professionalism.

It is also clear from the abundant case material covered in this volume that many people on all sides of large carnivore management are not happy with the way the institutional system currently operates. The need for an integrative and inclusive approach to large carnivore management is becoming increasingly obvious. Because existing institutional arrangements are so deeply woven into the fabric of society, the system as a whole is very conservative and not prone to change or adapt on its own. Its tendency to maintain the status quo ignores or hinders alternative ways of managing carnivores. Little in the world of carnivore management will change in the near future under present institutional arrangements.

Overcoming the institutional problem is possible in part by using a practice-based strategy, improving leadership, and creating a new metanarrative that is integrative and inclusive. These improvements mean working in the office and the field closely with ranchers, hunters, and environmentalists to address their concerns. The institutional system's new structure must view government agencies as partners, facilitators in clarifying and securing the common interest. The agencies and their professionals must behave in ways that make this possible. The foundation for a more effective institutional system has to be understanding and opening up dialogues through cooperative ventures that can be honestly appraised and refined. Many ongoing field projects could provide opportunities to develop participatory, common interest approaches, thus serving as a testing ground to distinguish what works, why, when, and where. The lessons can help us create more open and inclusive institutions so people will better listen to and address each other's expectations and demands. A reformed institutional system will stand a much greater chance of harmonizing different views into a broad-based consensus about managing large carnivores in the common interest. There is every reason to believe that we can achieve such a new arrangement.

References

1. This is not just a hypothetical example. Grizzlies have been expanding their range in recent years. See D. Darr, 2002, "Grizzly bear killed in Wyo. Range," *Daily Guide* (Jackson, WY), August 16–18, 1, 3; B. Farquhar, 2002, "Grizzlies heading south, southeast," *Casper Star-Tribune,* August 19, B 1; anonymous, 2002, "Grizzlies expanding range," *Jackson Hole Daily,* August 20, A9; S. Pyare, S. Cain, D. Moody, C. Schwartz, and J. Berger, 2004, "Carnivore re-colonisation: Reality, possibility and a non-equilibrium century for grizzly bears in the southern Yellowstone ecosystem," *Animal Conservation* 7, 1–7.

2. Much has been written on Wyoming, its history, and current dilemmas about natural resource management. An important recent summary is P. Krza, 2003, "Wyoming at a crossroads," *High Country News* 35(3), February 17, 1, 14–17. See also S. Western, 2002, *Pushed off the mountain, sold down the river,* Homestead Publishing, Moose, WY; D. Davy, 2002, *Cowboy culture: A saga of five centuries,* University of Kansas Press, Lawrence; P. N. Limerick, 1987, *The legacy of conquest: The unbroken past of the American West,* Norton, New York; R. Slotkin, 1973, *Regeneration through violence: The mythology of the American Frontier, 1600–1860,* Wesleyan University Press, Middletown, CT; M. Nie, 2003, *Beyond wolves: The politics of wolf recovery and management,* University of Minnesota Press, Minneapolis.

3. F. B. Samson and F. L. Knopf, 2001, "Archaic agencies, muddled missions, and conservation in the 21st century," *BioScience* 51, 869–873. Samson and Knopf's arguments are echoed in J. Berger and K. Berger, 2001, "Endangered species and the decline of American western legacy: What do changes in funding reflect?" *BioScience* 51, 591–593.

4. *Webster's New Universal Unabridged Dictionary,* 1994, Barnes and Noble, Avenel, NJ.

5. E. Ostrom, 1966, "Institutional rational choice: An assessment of the institutional analysis and development framework," in P. A. Sabatier, ed., *Theories of the policy process,* Westview Press, Boulder, CO. See also D. North, 1990, *Institutions, institutional change, and economic performance,* Cambridge University Press, Cambridge; H. D. Lasswell and M. S. McDougal, 1992, *Jurisprudence for a free society: Studies in law, science, and policy,* New Haven Press, New Haven; K. J. Meier, 1993, *Politics and the bureaucracy: Policy-making in the fourth branch of government,* Wadsworth, Delmont, CA; and V. C. Arnspiger, 1961, *Personality in social process: Values and strategies of individuals in a free*

society, Follett, Chicago. See also "Readings on institutional analysis: An overview," excerpt from the International Development Research Centre, http://web.idrc.ca/en/ev-3221-201-1-DO_TOPIC.html (accessed Feb. 1, 2004).

6. Ostrom, "Institutional rational choice: An assessment of the institutional analysis and development framework," 37.

7. H. J. Cortner, M. G. Wallace, S. Burke, and M. A. Moote, 1998, "Institutions matter: The need to address the institutional challenges of ecosystem management," *Landscape and Urban Planning* 40, 160; H. J. Cortner and M. A. Moote, 1999, *The politics of ecosystem management,* Island Press, Washington, D.C.

8. Perhaps the most universal conception of values is that of Harold Lasswell and Abraham Kaplan, 1950, *Power and society: A framework for political inquiry,* Yale University Press, New Haven. This approach provides a scientific way to study values and how they are shaped and shared through interpersonal interaction and through institutions. How the eight value categories can be used to analyze natural resource management cases is illustrated in T. W. Clark, 2002, *The policy process: A practical guide for natural resource professionals,* Yale University Press, New Haven.

9. V. C. Arnspiger, 1961, "Analysis and appraisal of social institutions" (ch. XI) and "Changing character of social institutions" (ch. XII) in *Personality in social process: Values and strategies of individuals in a free society,* Follett, Chicago. Three recent good examples of institutional analysis are C. B. Barrett, K. Brandon, C. Gibson, and H. J. Gjertsen, 2001, "Conserving tropical biodiversity amid weak institutions," *BioScience* 51, 497–502; M. A. Delmas, 2002, "The diffusion of environmental management standards in Europe and in the United States: An institutional perspective," *Policy Sciences* 35, 91–119; and S. Berstein, 2002, "International institutions and the framing of domestic policies: The Kyoto Protocol and Canada's response to climate change," *Policy Sciences* 35, 203–236.

10. T. Bridgman and D. Barry, 2002, "Regulation is evil: An application of narrative policy analysis to regulatory debate in New Zealand," *Policy Sciences* 35, 141–161, 141.

11. The structures and functioning of institutions of governance in natural resource management policy were the subject of R. D. Brunner et al., eds., 2002, *Finding common ground: Governance and natural resources in the American West,* Yale University Press, New Haven. This excellent book includes case histories of both wolves and grizzlies.

12. R. Hoskins, 2002, letter to Dr. Tom Thorne, acting director, WGFD, December 12. Hoskins cites I. N. Gabrielson, 1952, Report to the Wyoming Game and Fish Commission, Cheyenne, 31–32.

13. J. S. Dryzek, 1997, *The politics of the earth: Environmental discourses,* Oxford University Press, Oxford, 8. Dryzek offers a checklist of elements to analyze any discourse: (1) the basic entities recognized or constructed in the discourse, (2) the assumptions about relationships inherent in the discourse, (3) the agents that make up the discourse and their motives, and (4) the key metaphors (i.e., political symbols) and other rhetorical devices involved in the discourse and in advancing it over competing discourses. Formal study of these reveals how any discourse is structured, how it oper-

ates, and how competing discourses might work together to find common ground. Among Dryzek's other books on democracy are *Deliberative democracy and beyond: Liberals, critics, contestations,* 2000, Oxford University Press, New York; and *Discursive democracy: Politics, policy, and political science,* 1990, Cambridge University Press, Cambridge. Another relevant paper is R. G. Healy and W. Ascher, 1995, "Knowledge in the policy process: Incorporating new environmental information in natural resource policy making," *Policy Sciences* 28, 1–19. Many other authors have written about discourse analysis techniques.

14. S. Swaffield, 1998, "Contextual meanings in policy discourse: A case study of language use concerning resource policy in the New Zealand high country," *Policy Sciences* 31, 199–224.

15. C. MacDonald, 2003, "The value of discourse analysis as a methodological tool for understanding a land reform program," *Policy Sciences* 36, 15.

16. Analysis of claims and counterclaims provides insights into social process. See H. D. Lasswell, 1966, "The language of power," 3–19 in H. D. Lasswell et al., eds., *Language of politics,* MIT Press, Cambridge, MA; and Lasswell and McDougal, *Jurisprudence for a free society: Studies in law, science, and policy.*

17. Newspapers are full of claims and counterclaims about wolves and other large carnivores. For example, see D. Darr, 2002, "Wyoming wolf numbers reach nearly 200 in 2001," *Jackson Hole Guide,* April 15, A15; T. Morton, 2002, "Experts dispute wolf-kill claims," *Casper Star Tribune,* July 13, A1; W. Royster, 2002, "Forest wants clean camps this summer," *Jackson Hole News,* April 10, A3; W. Royster, 2002, "Wolf manager dispels myth about slaughter," *Jackson Hole News,* August 14, A29; J. Wilhelm, 2002, "Locals speak out against wolves in Sublette County," *Pinedale Roundup,* July 11, 11. See also R. T. Fanning, Jr., W. H. Hoppe, and D. Laubach, 2001, letter to President George W. Bush from Friends of the Northern Yellowstone Elk Herd, Inc., Pray, MT. T. Wilkinson, 2002, "Hatred toward wolves a sign of bigger things," *Jackson Hole News,* August 21, A6.

18. Darr, "Wyoming wolf numbers reach nearly 200 in 2001."

19. M. Rinehart, 2002, "Continue elk feeding, Wyoming sportsmen say," advertisement for Elk for Tomorrow, *Jackson Hole Guide,* April 3, A22.

20. M. Rinehart, 2002, "'Conservationists' ride public fear of brain-wasting disease to wipe out winter feeding of elk," advertisement by Elk for Tomorrow, *Jackson Hole Guide,* August 28, A26.

21. Fanning et al., letter to President Bush.

22. M. Phillips, 2002, Turner Endangered Species Fund, pers. comm.

23. A. M. Thuermer and W. Royster, 2002, "Enzi decries wolf," *Jackson Hole News,* July 17, A1, 18.

24. Rinehart, "'Conservationists' ride public fear of brain-wasting disease to wipe out winter feeding of elk"; Fanning et al., letter to President Bush; Thuermer and Royster, "Enzi decries wolf."

25. Incident analysis is an important research tool employed by W. M. Reisman and A. R. Willard, eds., 1988, *International incidents: The law that counts in world politics,*

Princeton University Press, Princeton. See also C. M. Cromley, 2000, "The killing of grizzly bear 209: Identifying norms for grizzly bear management," 173–220 in T. W. Clark et al., eds., *Foundations of natural resources management and policy,* Yale University Press, New Haven.

26. Reisman and Willard, eds., *International incidents: The law that counts in world politics,* 5.

27. Cromley, "The killing of grizzly bear 209: Identifying norms for grizzly bear management."

28. Both quotes are from Brunner et al., *Finding common ground: Governance and natural resources in the American West,* 21. Brunner cites C. J. Friedrich, taken from Lasswell and Kaplan, *Power and society: A framework for political inquiry,* 273.

29. Much has been written on "narrative" analysis. See J. Hukkinen, E. Roe, and G. I. Rochlin, 1990, "A salt on the land: A narrative analysis of the controversy over irrigation-related salinity and toxicity in California's San Joaquin Valley," *Policy Sciences* 23, 307–329. A recent study, Bridgman and Barry, "Regulation is evil: An application of narrative policy analysis to regulatory debate in New Zealand," is an excellent example. This section relies heavily on their study. See also M. Edelman, 1971, *Politics as symbolic action,* Markham, Chicago.

30. M. Stark, 2003, "Holes seen in Wyoming wolf plan," *Billings Gazette,* April 10; R. Huntington, 2003, "State quiets biologist who doubts wolf plan," *Jackson Hole News&Guide,* April 23, A17; T. Mangelsen, 2003, "Policies hurt speech, workers, wildlife," *Jackson Hole News&Guide,* April 23, A4; M. Bray and T. Mazzaarisi, 2003, "Governor fails first wildlife test," *Jackson Hole News&Guide,* April 23, A5; T. Wilkinson, 2003, "We need civil servants like Wyoming's Moody," *Jackson Hole News&Guide,* May 14, A6.

31. H. Harju, 2003, "Perspective: Politics is hurting wildlife management," *Casper Star Tribune,* April 14, A10.

32. Krza, "Wyoming at a crossroads."

33. Krza, "Wyoming at a crossroads."

34. Krza, "Wyoming at a crossroads."

35. Krza, "Wyoming at a crossroads."

36. J. Bruner, 1990, *Acts of meaning,* Harvard University Press, Cambridge, MA.

37. See, for example, R. B. Keiter and P. T. Holscher, 1990, "Wolf recovery under the Endangered Species Act: A study of contemporary federalism," *Public Land Law Review* 11, 19–52.

38. There are many good studies of bureaucratic behavior, including K. J. Meier, 1993, *Politics and the bureaucracy: Policymaking in the fourth branch of government,* Wadsworth, Belmont, CA; J. E. Gruber, 1987, *Controlling bureaucracies: Dilemmas in democratic governance,* University of California Press, Berkeley; J. Q. Wilson, 1989, *Bureaucracy: What government agencies do and why they do it,* Basic Books, New York; J. N. Clarke and D. McCool, 1985, *Staking out the terrain: Power differentials among natural resource management agencies,* State University of New York Press, Albany; C. W. Thomas, 2003, *Bureaucratic landscapes: Interagency cooperation and the preservation of biodiversity,* MIT

Press, Cambridge, MA; T. W. Clark, 1997, *Averting extinction: Reconstructing endangered species recovery,* Yale University Press, New Haven, looks at bureaucratic behavior in the case of Wyoming's endangered black-footed ferret.

39. See, for example, Wilson, *Bureaucracy: What government agencies do and why they do it.*

40. C. A. Perrow, 1979, *Complex organizations: A critical essay,* 2nd ed., Scott Foresman, Glenview, IL, 7.

41. See Clark, *Averting extinction: Reconstructing endangered species recovery,* especially ch. 8, "Problem definition: Analytical framework, process, and outcome," and related citations.

42. D. J. Decker and L. C. Chase, 1997, "Human dimensions of living with wildlife conservation," *Wildlife Society Bulletin* 28, 4–15; D. J. Decker, T. L. Brown, and G. F. Mattfeld, 1987, "Integrating social science into wildlife management: Barriers and limitations," 83–92 in M. L. Miller, R. P. Gale, and P. J. Brown, eds., *Social science in natural resource management systems,* Westview Press, Boulder, CO; S. J. Riley et al., 2002, "The essence of wildlife management," *Wildlife Society Bulletin* 30, 585–593; T. W. Clark and R. P. Reading, 1994, "A professional perspective: Improving problem solving, communication, and effectiveness," 351–369 in T. W. Clark, R. P. Reading, and A. L. Clarke, eds., 1994, *Endangered species recovery: Finding the lessons, improving the process,* Island Press, Washington, D.C.

43. See "Policy-oriented professionalism: A unique standpoint," in Clark, *The policy process: A practical guide for natural resource professionals;* "Civic professionalism: Meeting society's needs," 208–223 in Clark, *Averting extinction: Reconstructing endangered species recovery;* T. W. Clark, M. B. Rutherford, K. Ziegelmayer, and M. J. Stevenson, 2001, "Conclusion: Knowledge and skills for professional practice," 253–276 in T. W. Clark et al., *Species and ecosystem conservation: An interdisciplinary approach,* Yale School of Forestry and Environmental Studies Bulletin No. 105, http://www.yale.edu/environment/publications.

44. The new paradigm for professionalism is being discussed in many scientific and professional circles. A small sampling of literature includes "Civic professionalism: Meeting society's needs," 208–223 in Clark, *Averting extinction: Reconstructing endangered species recovery;* D. A. Schön, 1983, *The reflective practitioner: How professionals think in action,* Basic Books, New York; T. W. Clark, 1997, "Conservation biologists: Learning to be effective and practical," 575–597 in G. K. Meffe and C.R. Carroll, eds., *Principles of conservation biology,* Sinauer Associates, Sunderland, MA.

45. For a description of the practice-based approach for improving management, see R. D. Brunner and T. W. Clark, 1997, "A practice-based approach to ecosystem management," *Conservation Biology* 11, 48–58; Brunner et al., *Finding common ground: Governance and natural resources in the American West;* T. W. Clark et al., 2000, "Koala conservation policy process: Appraisal and recommendations," *Conservation Biology* 14, 681–690; T. W. Clark and J. Padwe, 2004, "The Ecuadorian Condor Bioreserve initiative: Decision process considerations for effective conservation," *Journal of Sustainable Forestry,* 18, 297–324.

46. See also P. Sowka and M. Madel, 2003, Living with wildlife program, Psowka@blackfoot.net.

47. T. Kaminski, C. Mamo, and C. Callaghan, 2002, "Managing wolves and bears where private ranches meet public reserves," *NRCC News* (Northern Rockies Conservation Cooperative, Jackson, WY) 15, 11–12, 16; C. Callaghan, 2003, "Crossing borders: The many challenges of trans-boundary wolf conservation," *Howlings* 12(1), 1, 6–7.

48. S. Primm and L. Lasley, 2001, "Piecemeal plans best for grizzly delisting," *Jackson Hole News,* April 4, A7.

49. S. Primm, 2002, "Grizzly conservation and adaptive learning," *NRCC News* 15, 7–8.

50. S. M. Wilson, 2003, "Conservation on the edge: Lessons for recovering grizzly bears in contested landscapes," *NRCC News* 16, 3–5.

51. S. R. Brown and T. W. Clark, 2003, "Workshop on large carnivore conservation," *NRCC News* 16, 12, 16; S. R. Brown, 2003, "Workshop on large carnivore conservation," (summary report to participants), Department of Political Science, Kent State University, Kent, Ohio.

52. Much has been written about leadership. See, for example, J. M. Burns, 1978, *Leadership,* Harper, New York; D. A. Schön, 1987, *Educating the reflective practitioner,* Jossey-Bass, San Francisco; A. Brandenburg and A. Boelter, 2000, *Assessment of collaboration and consensus building needs and opportunities in Wyoming,* Institute and School for Environment and Natural Resources, University of Wyoming, Laramie.

53. R. Sidaway, 2003, "*Good practice in rural development,*" Consensus Building, Central Research Unit, Saughton House, Broomhouse Drive, Edinburgh, Scotland. See also A. Flores and T. W. Clark, 2001, "Finding common ground in biological conservation: Beyond the anthropocentric vs. biocentric controversy," 241–252 in Clark et al., *Species and ecosystem conservation: An interdisciplinary approach.*

54. E. Roe, 1994, *Narrative policy analysis: Theory and practice,* Duke University Press, Durham, NC. See also E. Roe, 2001, "Varieties of issue incompleteness and coordination: An example from ecosystem management," *Policy Sciences* 34, 111–133.

Chapter **8**

Coexisting with Large Carnivores: Lessons from Greater Yellowstone

Murray B. Rutherford and Tim W. Clark

Large carnivores test our commitment to conservation. They require large areas of habitat for survival. They compete with us as top predators. They occasionally interfere with our valued activities, such as raising livestock and keeping domestic pets. Large carnivores can also be frightening, and in rare cases they are actually dangerous to humans.

Not only do large carnivores test our commitment to conservation, but our efforts to coexist with these animals severely challenge our ability to make and implement wise decisions. As the case studies in this volume show, individual and collective choices about mountain lions, grizzly bears, and wolves have often advanced narrow special interests at the expense of carnivores and the broader common good. Whether it has been the interests of the hunting community driving decisions about mountain lions, or the demands of ranchers dictating decisions about grizzly bears and wolves, or the professional biases and technical scientific expertise of wildlife professionals dominating all management processes, or the endless power contest between the state and federal governments, the voices of special interests have frequently been given undue weight in decision making. Too often, the institutions of wildlife management have failed to engage a broad range of participants to identify, select, and implement policies and practices that advance the common interest in an enduring way.

In chapter 1 we describe the overall goal of large carnivore management as "restoring large carnivores and carrying out programs for coexisting with them in ways that will engage the public and benefit from public support." There is little doubt that this goal represents the broader common interest. Evidence can be found at the international level in treaties such as the Convention on Biological Diversity, at the national level in legislation such as the Endangered Species Act, and even at the regional and local levels in the justifications for many species management plans and other conservation policies. As we state in chapter 1, this goal "is likely to be supported by all or most people, at least in principle."

In practice, however, participants in wildlife management have had difficulty sorting out what this common interest goal means in local contexts and how it can be achieved on the ground. Also, too often power has been used by some participants to override the views of others about appropriate goals and how to achieve them. Conflicting interests have repeatedly clashed over management decisions, without satisfactory processes to encourage constructive dialogue and resolution of disputes. Over time, this has led to a highly politicized decision-making context, in which the dominant motivation among many participants is maximizing their own power to influence choices.

The authors in this volume argue that these management weaknesses are deeply significant, not only for the possibility of coexisting with large carnivores in Greater Yellowstone, but for managing wildlife more generally, and for maintaining civil society and effective systems of democratic governance. Given current trends of population growth and habitat loss throughout the world, it is crucial for the survival of species and ecosystems that humans learn to live with wildlife outside protected areas. It is also crucial that our systems of governance find ways to move beyond intractable conflicts and gridlock.

Fortunately, there are positive as well as negative lessons about governance and coexistence to be learned from experiences in Greater Yellowstone. In this concluding chapter, we review the lessons that can be gathered from the studies in this book, summarize key steps needed to move toward coexistence in this region, and discuss the implications for management settings elsewhere.

Lessons from Greater Yellowstone

Perhaps the primary message of this volume is summarized in chapter 2: "Managing large carnivores at present is less a problem of animal biology or control and more a problem of cultural perception, decision process, and institutional dilemma." People and their decision making are by far the most significant cause of mortality for large carnivores. Although each of the case studies summarizes the biology and behavioral ecology of the species and discusses the implications of these factors for management, the focus throughout is on the human side of the coexistence problem. In particular, the studies highlight the importance of understanding the context for carnivore management, reducing the frequency and impact of actual carnivore-human conflicts, and addressing pervasive weaknesses of governance.

Understanding the Context for Carnivore Management

According to policy scientist Ron Brunner, "Most *preventable* errors of policy analysis stem from the analyst's perspective. As the analyst simplifies a problem to make it tractable for analysis and action, some important part of the relevant context is misconstrued or overlooked altogether. The analytical error—what is misconstrued or overlooked —becomes apparent only in retrospect, after resources have been committed and the unintended and often adverse results start coming in."[1] Preventable analytical errors of the type that Brunner described have occurred repeatedly in large carnivore management. Chapter 3 discusses the unintended and adverse results of the decision by the Wyoming Game and Fish Department (WGFD) to increase the mountain lion hunting quota in the region of Jackson in spite of a recent rise in public empathy for lions generated by a highly visible lion family denning nearby. The authors argue that WGFD did not pay sufficient attention to this change in the context. Chapter 4 notes that the survival of grizzly bears depends on the behavior of individual people as they participate in recreational activities or go about earning their livelihoods. This behavior cannot be monitored continuously or controlled by force, so a thorough understanding of the perspectives and motivations of these individuals is essential to good decision making about bear management. Again, these are matters of context. Chapter 5 describes how the compensation program for ranchers that suffer livestock losses as a result of wolf depredations has been criticized for not being sufficiently flexible to adjust to the day-to-day circumstances of different ranchers, for not providing ranchers with the right incentives to change husbandry practices, and for not compensating for other types of losses caused by the presence of wolves, including the deprivations of respect and skill that ranchers feel when they lose livestock to wolves. These are all features of the context.

In short, *context matters,* and it is just as important for large carnivore management as it is for any other kind of decision making. Mapping and understanding the context is the first step toward wiser decisions, and an entire chapter of this book is devoted to this task. In chapter 2 the authors lay out a detailed map of the participants, their perspectives, the situations in which they interact, their sources of power, the strategies they use to achieve their goals, and the outcomes and ultimate effects of those social interactions, including the decision processes that have evolved. In doing so, they not only describe the context for large carnivore management in this setting, but also demonstrate the use of a framework for contextual mapping that can be applied in other management settings.[2] Their map is preliminary and open to refinement with further experience and additional data, but it highlights several key features of the context that have been particularly problematic for coexistence: deeply entrenched conflict among participants, highly symbolic politics, increasing numbers of carnivore-human incidents, and weak institutional response. These features of the large carnivore management context are emphasized throughout this volume.

According to the authors of chapter 2, the seemingly intractable conflict over large carnivore management is driven by strongly polarized worldviews, or perspectives. Perspectives are made up of values demanded from life, attitudes toward nature, self-identifications, beliefs about how the world works and where it is going, and symbolic support for these identifications and beliefs. At the risk of oversimplifying, the main perspectives at play can usefully be categorized into three groups. First is the *localist* perspective—people who are interested in using and dominating nature for human purposes and who tend to identify with the myths, culture, and symbols of the Old West. The second type we have called the *environmentalist* perspective—those with strong moral and ethical concerns for nature, who identify with the myths, culture, and symbols of the New West. And the third perspective is that of *agency personnel,* who may agree either with localists or environmentalists in many of their personal views, but who share with each other an identification with professional and bureaucratic norms and practices. Each set of participants feels that its own views are correct and should prevail, including its views about who should have power in decision making.

The contest between the belief systems of the Old West and the New West is especially significant, involving different understandings of the relative importance of individual versus collective rights, local versus national interests, the state versus the federal government, and agriculture versus other livelihoods. Many aspects of this debate can be traced to historical struggles about the appropriate structure of democracy at the time the nation was formed—struggles that remain to some degree unresolved.

In the last few decades, carnivore management in the West has become linked to these broader political struggles in many people's minds. One plausible explanation for this is that the forces for reintroduction and conservation have been strongest at the national level. These forces have been expressed mainly through national policies, whereas the forces of localism and carnivore control have been, as the name "localist" itself implies, concentrated at the state and local levels and have been expressed through policies at these levels. Localists feel that carnivore conservation initiatives are being forced upon them by outside forces over which they have little control, whereas environmentalists feel that justified national policies are being thwarted by unreasonable and self-oriented parochial interests.

One clear sign that such deeper political factors are at play is the highly emotional nature of the claims and counterclaims made about carnivores and carnivore management. Large carnivores have always been emotionally charged symbols, but the level of emotion has become heightened in recent decades, and the symbols have become more dissociated from real meanings. Scientific data and facts are often drowned in symbolic struggles.

The problems of deeply entrenched conflict and highly symbolic politics have been maintained and exacerbated in the last few decades by two other key contextual features: the increasing incidence of actual carnivore-human conflicts, and the inability of current governance structures to ameliorate conflict and find lasting common interest solutions. These features are sufficiently important to receive separate treatment in the next two subsections of this chapter.

Real Carnivores and Real Conflicts

Although much of the debate about large carnivore management is shaped by symbolic understandings rather than solid data, there are real carnivores behind the symbolism, and actual conflicts do occur between these animals and people. The studies in this volume show how the frequency of such conflicts outside of protected areas has increased in recent years. Several trends can be identified as contributing to this increase. First, human populations are growing rapidly in the West, as new residents move in and large landholdings are subdivided to form ranchettes and accommodate trophy houses. The increase in numbers of humans and the associated development of private lands at the interface with public lands increase the probability of conflict with carnivores. At the same time, recreational use of public lands is increasing, and this also potentially brings more people into contact with carnivores. Finally, the number of wolves and grizzly bears outside protected areas has been increasing because of the success of reintroductions and conservation efforts. Mountain lion numbers may have increased as well in the last decade, although there are not sufficient data to be certain. As the populations of wolves and bears within protected areas have grown, these animals have dispersed into the adjoining landscape, into areas that are intensively used for raising livestock or are occupied by the expanding human population. This brings carnivores into conflict with traditional land use practices.

Given the trends of growth in human and carnivore populations, it was inevitable that there would be more contact and conflict between humans and animals outside protected areas. Mountain lion sightings have gone up since the early 1990s. Incidents of grizzly bear predation on livestock and interference with other human activities have increased dramatically during the same period. Wolves have killed sheep and cattle. Although these trends were quite predictable, there has been a tendency to react to incidents after the fact rather than taking proactive steps to minimize and mitigate foreseeable conflicts.

Several of the authors in this volume point out, however, that in spite of these increases in carnivore-human conflicts, large carnivores still account for only a small proportion of livestock losses when compared to other causes, and the extent and consequences of large carnivore predation on ungulate populations are contested. Moreover, ranching actually forms only a small and decreasing part of the economic base of the region, and human injury or death from carnivores is still very rare. But the symbolic linkage of large carnivores to larger political issues allows each incident of conflict—whether it is a grizzly bear killing sheep, a mountain lion taking pets, or wolves killing cattle—to be inflated and used to fuel the ongoing politicization of the management process. Perceptions are what counts, and each incident reinforces perceptions of vicious predators foisted on localists by outside forces and the continuing decline of the cowboy way of life and the Old West.

Finding ways to diminish the frequency and magnitude of actual carnivore-human conflicts, then, is essential. As is argued in chapter 4, horror stories about such conflicts may be used by anticonservation forces to justify attacks at the national level on conservation programs, budgets, and legislation. On the other hand, proconservation forces may use the killing of depredating animals, which is the normal outcome of such incidents, to call for more draconian, top-down prescriptions, which will in turn magnify rather than relieve the conflict.

Failures of Governance

Much of this book is about failures of governance. Each of the case studies describes the increased politicization of management of the species under consideration. Recall that politicization refers to the process whereby, because of continuous and unresolved struggles and deeply entrenched conflict among participants, trust erodes and the will to search for common interest solutions is lost, together with any faith that such solutions exist. In these circumstances, power becomes all-important—participants see their interactions with others as battles, and they seek power above all else to advance their own interests. Politicization takes place when systems of governance are incapable of working through conflict to find socially acceptable outcomes.[3]

Good governance allows interested parties a real voice in decision making, incorporates a wide range of viewpoints, proactively keeps conflict within reasonable bounds, and seeks common ground solutions to social problems. In large carnivore management, governance has fallen far short of these ideals. For example, chapter 3 describes how WGFD has embodied localist values, allowed little public input into its decisions, and failed to specify and adopt clear goals for mountain lion management, thereby confounding conflicts with its goals for the management of other species. Chapter 4 discusses the failure of the U.S. Fish and Wildlife Service to establish clear public expectations about the meaning of the "recovery zone" for grizzly bears or to work with residents outside that zone to prepare for grizzly bear dispersal. This has set up conflict between those who believe that bears should be restricted to the recovery zone and those who believe that this zone is just the first step in a recovery effort that will extend well beyond its boundaries. Chapter 5 argues that WGFD is headed for the same problems with wolf management that it has encountered with grizzly management, as it positions itself to take over primary responsibility for wolves from the Fish and Wildlife Service. The authors note the agency's inability to adopt clear goals for wolf management, the real possibility of conflict with its other mandates and its existing funding sources, the linkages of the agency with states' rights and localist views, and the general lack of agency resources for wolf management. Of particular concern is the state's decision to classify wolves as "trophy game" in national forest wilderness areas that adjoin Yellowstone National Park and as "predators" in all other areas of the state, thus subjecting almost all wolves in the state to unlimited hunting as they travel outside protected areas.

As the management processes for wolves and grizzly bears move toward delisting and devolution of management to the state level, there should be great concern about these questions of governance. Federal agencies have defined management success on the basis of specific, short-term population goals, but they do not seem to have adequately addressed all the long-term uncertainties of persistence, including many of the conditions that caused population declines outside protected areas in the first place.[4] State agencies, meanwhile, have not demonstrated that they have the capacity to manage wolves and grizzly bears for long-term persistence, and some observers question whether they are even committed to this goal.

Two chapters of this volume are devoted entirely to improving governance. In chapter 6, the authors explore participatory approaches to decision making. They argue that civil society is deteriorating in Wyoming and elsewhere in the United States because "people are failing to resolve their collective disputes or to establish an agreeable process for redressing grievances." Ongoing conflict has become institutionalized, built upon repeated, dysfunctional patterns of interaction, eroded trust, and an increasing focus on power. In this environment participants naturally gravitate toward dichotomous, opposing points of view, "taking sides" by defining themselves in opposition to the views of their perceived opponents. Meanwhile, agencies struggle to reconcile their conflicting mandates, such as multiple use, sustained yield, and endangered species protection. The authors of chapter 6 argue that "the key to breaking these dysfunctional patterns of interaction is to get local participants working together on real, manageable, on-the-ground problems where power and control are not such major issues and symbolic debate is minimized."

Chapter 7 critiques the institutional system of wildlife management. Institutions are formed to pursue desired human values; democratic institutions promote the production and broad distribution of all desired values across society. Our wildlife management institutions, however, have been unable to resolve conflicting claims and counterclaims about carnivore management or to find integrative bridges across competing discourses. The authors go on to deconstruct the main competing narratives about carnivore management. The dominant narrative is about individualism, the Wild West, and carnivores as unacceptable risks to life and livelihood imposed by outside forces on localists, who must live with the negative consequences. This story is told by the localists themselves and by many state officials. The principal counter narrative is a story told by environmentalists and some federal representatives, about what they perceive to be the best interests of the nation as a whole, carnivore conservation as the will of the nation, as they see it, and the "unreasonable localists" who are putting their own selfish interests ahead of the national good. Much of the debate over large carnivores is about which of these narratives should be embedded in the institutional system of wildlife management. At present state and local institutions largely embody the dominant narrative of the localists.

The authors argue that these institutional problems are compounded by bureaucratic structures and professional operating norms of the wildlife management agencies and their staff. Bureaucracies tend to rely on hierarchical structures of authority and control, standard operating procedures, and traditional approaches to problems. They vigorously defend their power bases and refuse to share power with the public, and they deal poorly with complex, uncertain, and rapidly changing environments. None of these attributes is conducive to good governance in the context that exists for managing large carnivores. Rather than finding integrative solutions, the bureaucracies of wildlife management contribute to the ongoing divisions. Professional norms include the assumption that biological data are sufficient for good management and the assumption that professional training in the natural sciences is sufficient to develop good managers. Accordingly, the professional's argument goes, these managers should be given privileged positions in decision making because of their expertise. In fact, however, these assumptions are not actually supported by theory or practice.

Moving toward Coexistence

How can we move toward sustainable coexistence with large carnivores? Three main themes run through the recommendations of all of the authors in this volume: (1) working from the bottom up rather than the top down, engaging local participants in crafting innovative, contextually appropriate, practice-based prototypes and learning adaptively from experience; (2) changing the meanings of carnivores and the symbolic linkages of these animals with broader political issues; and (3) restructuring and upgrading the capacity of the institutional system of wildlife management to incorporate a wider range of values and resolve conflict by seeking and implementing common interest solutions.

Locally Designed Solutions

Large carnivore management takes place in highly complex, uncertain, and locally variable contexts. It is characterized by symbolic politics and politicized management processes. Coexistence will require substantial changes in the behavior of individual humans in situations that are difficult to monitor and control. Many of the individuals whose behavior is most significant are highly resistant to federal regulation and activity. All of these factors support a strategy of developing localized, contextually appropriate innovations for living with carnivores, rather than sweeping, one-size-fits-all programs imposed from above. Given the present conditions of political gridlock at the federal level, it is unlikely that new national programs for carnivore conservation will be prescribed in any event, and even if prescribed, they are unlikely to be implementable at the local level. We summarize the justifications and preferred design elements for local, practice-based programs here.

The first step discussed by many of the authors in this volume is to identify those areas where the frequency of human-carnivore incidents is now high or is likely to increase substantially in the near future. Chapter 6 describes how local, citizen-driven, participatory projects could be developed in these areas to engage citizens in working together to resolve on-the-ground conflicts, thereby building trust and social capital and diffusing symbolic debate. Many opportunities exist for such interventions. For instance, in chapter 3 the authors suggest that initial participatory projects for mountain lion management could focus on the shared goal of finding ways locally to reduce the number of lions that are removed for getting in trouble with human activities. These projects might include eliminating winter deer feeding and other activities that are risky in mountain lion country. Over time groups that begin by working with such issues might be able to graduate to making broader recommendations for a formal, regional management plan for mountain lions.

Chapter 4 emphasizes the importance of dialogue, or reasoning together in local, participatory, decision-making processes. The authors suggest that participatory processes concerning grizzly bears might be initiated by a respected nongovernmental organization, which could convene local meetings to discuss and implement projects such as improving food storage and sanitation or training programs to teach campers, hikers, hunters, and homeowners about bear-safe behavior. From these modest beginnings, local processes could work toward developing funding programs for grizzly bear conservation, designing better compensation mechanisms for depredations (e.g., insurance pools), and ultimately making recommendations for regional management plans and programs. For wolf management, the authors of chapter 5 argue that "a hands-on, cooperative, evidence-based approach will minimize carnivore conflicts while at the same time ameliorate some harmful political dynamics." They also suggest small-scale projects, such as identifying and working with ranchers who are interested in incorporating safe husbandry practices and working together with agency representatives to institute more flexible grazing practices that are less susceptible to wolf depredation.

As these examples show, in addition to designing projects to reduce the damage caused by carnivores directly, local processes can begin to develop collectively reasoned opinions about broader strategies to deal with carnivore management problems. As the authors of chapter 6 observe, "The promise of participatory problem solving lies in its potential to find out what a spectrum of thoughtful and engaged people think about the nature of a particular problem and how we should go about solving it." However, it is important to begin with less ambitious goals, spending time together dealing with real problems, to build trust and capacity before advancing to more contentious undertakings.

Another important early task for local initiatives is to develop better understanding among participants of the context in which they are operating, including their own identities, value demands and expectations, and those of other participants. This involves turning attention to mapping the context, using techniques such as those outlined in chapter 2. It also involves gathering data—not just biological data about mountain lions, grizzly bears, and wolves—but data on people's perspectives and the mechanisms through which decisions are made. Techniques for exploring subjective worldviews that encourage self-reflection, such as Q method, can be useful here.[5] As a contextual map is developed, it can be shared with others in the community to encourage all participants to improve their understandings of self-in-context.

To learn from experience, such local efforts need to be integrated with an effective appraisal program, which regularly and independently assesses progress, harvests the lessons (good and bad), communicates those lessons to other settings, and encourages adoption and adaptation of the ideas that have been successful. Appraisal and diffusion of lessons should not be limited to new initiatives. A variety of innovative local programs already exists that can serve as models and inspirations for other settings.

Changing the Symbolic Meaning of Carnivores

One of the most effective ways to move away from contentious abstract debate about the symbolic political meanings of carnivores is to deal with real problems caused by real carnivores on the ground. Small-scale, local projects such as those described above can begin to substitute actual experiences with carnivores for exaggerated myths and to unlink carnivore management from larger political debates.

Also, by reducing the frequency and magnitude of actual impacts, local projects would limit the ammunition available for manipulation by special interests.

Chapter 6 recommends broad-based, participatory outreach activities to promote new meanings for large carnivores. The authors suggest that outreach activities should evolve through three main phases: an engagement phase, in which participants with diverse views are asked to discuss their perspectives and explore potential common ground; a collaboration phase in which participants work together toward redefining the problems and meanings of carnivores; and a formalization phase, in which participants commit to new approaches.

Through such efforts, large carnivores may over time become more accepted as natural conditions of the local environment and perhaps even as sources of pride. As the authors outline in chapter 7, this may lead to the development of a new, more integrative meta-narrative about large carnivores and their management—perhaps a story of cooperative problem solving and the ability of modern westerners to coexist with fierce but magnificent predators.

Restructuring the Institutions of Wildlife Management

Decisions about managing and coexisting with large carnivores are made through the institutional system of wildlife management. Chapter 7 details current problems with the structure and function of this system, and the case studies provide numerous examples of institutional dysfunction. The strategies described above to develop practice-based projects and change the symbolic meaning of large carnivores necessarily involve some degree of institutional change. Because institutions are built from beliefs and practices over time, these strategies should also lead in the longer term to more substantial institutional restructuring. For example, the recommendation in chapter 3 for participatory development of a new regional lion management plan represents a small-scale project that builds toward more substantial institutional change. Similarly, chapter 6 suggests that, after a period of time, the groups involved with participatory projects might achieve sufficient levels of trust and understanding that they could develop alternative conservation plans, appoint delegates to regional carnivore panels, and eventually build new cooperative community institutions.

Institutions, however, are stubbornly resistant to change. As systems scientists Lance Gunderson, C. S. Holling, and Steve Light observed about government agencies: "The ability of the bureaucracy of a government agency to control information and resist change seems to show a level of individual and group ingenuity and persistence that reflects conscious control by dedicated and intelligent individuals as well as the unconscious part of the organization and culture of bureaucracies. Collaterally, the culture of organizations imposes unconscious constraints."[6] Additional measures will be needed to create the institutional space for innovation in large carnivore management and to kick start the dramatic institutional reform that is needed. The authors of chapter 7 recommend steps to establish and fund improved institutional appraisal capacity and to open lines of communication among local, regional, and state initiatives in order to share experience and expertise, diffuse lessons, and encourage learning. They also argue that training highly skilled, problem-oriented leaders with expertise in the policy sciences to complement existing expertise in the natural sciences is a key element in building a better institutional system of wildlife management.

Applying Lessons in Other Settings

What do these experiences in Greater Yellowstone tell us that is most relevant for management settings elsewhere? First, they tell us much about the problems that humans encounter in attempting to live with large carnivores outside protected areas and how to intervene to resolve these problems. Second, they suggest how the traditional relationship between professional wildlife management experts and the public often fails to serve the common interest and how this relationship might be restructured. Third, they provide clues for designing governance processes that are better suited to the conditions of modern social and ecological systems. We expand on each of these points below.

In many places in the world, the regions adjacent to protected areas have become hot spots of conflict between animals and humans.[7] If such conflicts and the resulting wildlife mortalities could be reduced, these regions would provide important additional habitat and linkages among protected areas. Minimizing conflicts and mortalities will require either a constant paramilitary presence to control the actions of humans, which is unpalatable and in most cases infeasible, or working with local residents to understand their perspectives and practices and to learn ways of coexisting. The strategies proposed in this volume—designed to reduce carnivore damage, change symbolic meanings of carnivores, and build local initiatives through which participants clarify and pursue their common interests—are geared toward learning ways of coexisting. These strategies may be successful in other settings as well as in the Yellowstone region, but the key will be to map the context for management, adapt the strategies to fit the setting, monitor progress, and adjust or terminate as indicated by the results.

The tension between professional wildlife management experts and the public also exists throughout the world because professional, scientific norms are widely shared. There is a global need to move toward a more civic-minded model of professional practice in wildlife management, in which democratic processes and the common interest are given due attention and sources of knowledge other than positivistic methods of scientific inquiry are given due weight in decision making.[8] The recommendations by the authors of this book for pragmatic professionalism and leadership training would help the progression toward that model.

Finally, Greater Yellowstone also has broader lessons to offer about appropriate governance. Modern society has changed dramatically, as has our knowledge of ecosystems, and management institutions and means of democratic governance need to adapt to these changes. Rigid, bureaucratic agencies and hierarchical, top-down, governance structures have great difficulty dealing with nonlinear, uncertain, and unpredictable ecosystems. Moreover, they have great difficulty dealing with increasingly complex, diverse, and interest-based modern social systems.[9] The type of locally based initiative discussed in this volume offers a means to redevelop social capacity and governance structures at the community level that are more flexible, adaptable, and capable of learning.

Conclusion

Finding ways to coexist with large carnivores is a difficult, but not impossible, challenge. Case studies of mountain lions, grizzly bears, and wolves indicate that we are having great trouble meeting this challenge in Greater Yellowstone. If we do not find ways to do better, then populations of large carnivores in the protected areas of this region will simply act as sources for mortality sinks immediately beyond the margins of these protected areas. Such a condition seems highly unlikely to ensure long-term sustainability, given climate change, invasive species, disease, and other factors that may reduce the effectiveness of existing protected areas. The research discussed in this book shows that failure to understand the sociopolitical context for management initiatives, symbol inflation and linkages with highly contested political issues, and institutions that are unable to work with multiple interests to find and secure the common interest are all key problems in our attempts to coexist sustainably with large carnivores in this region. We recommend mapping the context and working with local people on local problems to reduce conflict, change the symbolism, and initiate institutional reform.

This is a crucial time for large carnivore management in the western United States. Through delisting, management of wolves and grizzly bears is on the verge of moving from a federally controlled, Endangered Species Act–driven regime to a state-managed process. Some hope that this will move management out of the spotlight of national public scrutiny and controversy and will allow a return to the ways of the past. But the context for large carnivore management has changed dramatically and the ways of the past will no longer satisfy or suffice.

There is a wonderful opportunity in Greater Yellowstone to model sustainable coexistence and progressive governance for ourselves and for the world. We will need to use the best available ideas, knowledge, and leadership to get there. It is our hope that this book will stimulate constructive discussion about the problems of coexistence and the proactive interventions that will be required to resolve these problems.

References

1. R. D. Brunner, 1991, "The policy movement as a policy problem," *Policy Sciences* 24, 65–98, 65.

2. The concept and methods of contextual mapping are discussed by H. D. Lasswell, 1971, *A pre-view of policy sciences,* American Elsevier, New York; this book laid the groundwork for this kind of mapping. See also T. W. Clark, 2002, "Social process: Mapping the context," 32–55 in *The policy process: A practical guide for natural resource professionals,* Yale University Press, New Haven; T. W. Clark, 2001, "Interdisciplinary

problem solving in species and ecosystem conservation," 35–54 in T. W. Clark, M. Stevenson, K. Ziegelmayer, and M. Rutherford, eds., *Species and ecosystem conservation: An interdisciplinary approach,* Yale University School of Forestry and Environmental Studies Bulletin 105.

3. H. D. Lasswell and M. S. McDougal, 1992, *Jurisprudence for a free society: Studies in law, science, and policy,* 2 vols., New Haven Press, New Haven.

4. T. W. Clark and A. M. Gillesberg, 2001, "Lessons from wolf restoration in Greater Yellowstone," 135–149 in V. A. Sharp, B. Norton, and S. Donnelly, eds., *Wolves and human communities: Biology, politics, and ethics,* Island Press, Washington, D.C.

5. See K. Bryd, 2002, "Mirrors and metaphors: Contemporary narratives of the wolf in Minnesota," *Ethics, Place, and Environment* 5, 50–65, and citations therein on Q method.

6. L. H. Gunderson, C. S. Holling, and S. S. Light, 1995, "Barriers broken and bridges built: A synthesis," 489–532 in L. H. Gunderson, C. S. Holling, and S. S. Light, eds., 1995, *Barriers and bridges to the renewal of ecosystems and institutions,* Columbia University Press, New York, 517.

7. J. L. Gittleman, S. M. Funk, D. W. MacDonald, and R. K. Wayne, eds., 2001, *Carnivore conservation,* Cambridge University Press, London.

8. T. W. Clark, 1997, "Civic professionalism: Meeting society's needs," 208–223 in T. W. Clark, *Averting extinction: Reconstructing endangered species recovery,* Yale University Press, New Haven.

9. R. D. Brunner, C. H. Colburn, C. M. Cromley, R. A. Klein, and E. A. Olson, eds., 2002, *Finding common ground: Governance and natural resources in the American West,* Yale University Press, New Haven.

Appendix

Making Carnivore Management Programs More Effective: A Guide for Decision Making

Tim W. Clark and David J. Mattson

Proposed here is a comprehensive series of questions, based on the text and lessons of this volume, that can help people think constructively about organizing and making decisions in a large carnivore conservation program. The questions are designed to help everyone carry out successful programs, whether they are new programs that are being set up or existing ones that are facing conflict or undergoing review.

Answering these questions may appear to be an academic or theoretical exercise, but the questions are systematically presented to help people be as rational and practical in their work as possible. There are no single, right answers, of course. The purpose of this exercise is to encourage people to be deliberate, systematic, and thorough in examining themselves, the structure and functioning of the program, and the process of decision making. This exercise is applicable to a broad range of cases, and these questions are appropriate for scientists, managers, decision makers, citizens, researchers, investigative reporters, ranchers, advocates, and anyone else affected in any way by a large carnivore management program. Discussing and comparing the diverse answers that result is a good way to begin building trust so that participants can delve into the causes and conditions of their problems and, ultimately, explore alternatives in a creative and inclusive way. This exercise can help people find ways to identify common interests, clarify their goals, and track progress toward achieving those goals. It can also help them make the kinds of adjustments needed to make the program more successful.

We encourage users of this guide to modify or adapt the questions to fit their particular situation. You can refer back to the text to help you think about and use these questions.

I. How well are the overall conservation program and its decision-making process working?

1) How would you characterize an ideal conservation program? What features would it have? Be specific.
2) How does the current program function? Describe its strengths and weaknesses.
3) What are the differences between 1 and 2? Again, be specific.
4) Explain the differences. In other words, what factors are causing or contributing to differences between the ideal program and the actual program? Such factors might include a lack of clear or realistic goals, the wrong program structure, weak leadership, lack of skills on the part of the professionals involved, the wrong equipment, or too few resources. Try to explain the program's functioning in terms of the people involved, the decision-making process, and similar "systems" variables.
5) Identify possible means that participants could use to minimize these differences or address the problems that you have identified. That is, how can you, as a group, move the current effort toward the ideal and close the gap between how the program currently operates and how it should operate? Be creative. Don't settle on the first idea that comes to mind, but let the group spend a long time fully discussing and evaluating lots of ideas. To answer this question, you need to refer to 4 above. The suggested alternatives should be geared specifically to the variables or problems you identified that must be changed to improve the program (e.g., change goals, decision-making process, leaders, or some other variable). Evaluate each suggestion realistically. Explicitly, how will each alternative solve the program's problems?
6) Which of the suggestions or alternatives developed under 5 above are most promising? Ask which problems will be solved by which suggested change. Will the proposed changes improve the pro-

gram, or might they create other, unintended consequences? How will you measure progress if you implement the suggestions?

II. How well are the involved people, groups, and organizations interacting with one another in the existing program?

Get data on these matters; do not rely on casual opinion. Informed observation is important in answering all these questions.

1) Who are the key participants (individuals and organizations, official or unofficial) in the program? Who is not participating? Who should be involved?
2) What are the perspectives, goals, assumptions, expectations and values of participants in the program? Getting information to answer this question can be difficult, but it is important.
3) In what situations or settings do participants interact (science, management, media, courts, other)? Is there a way to change the patterns of interaction for the better?
4) What strategies do participants favor or use to get their way? In an open democracy, persuasive strategies are more sustainable and often more effective than coercive ones. You can use education, diplomacy, or economics in a persuasive way. Sometimes coercion seems justified, but it is often destructive in the long run.
5) What are the short-term outcomes and long-term effects of these interactions on the people involved in the program, on public perceptions, on management institutions and decision-making processes, and on carnivores and their habitats? These are important questions: it is possible to save carnivores in the short term, but alienate the public, make institutions more rigid or defensive, or create other problems so that long-term conservation becomes impossible.

III. How well are decisions being made?

This set of questions clarifies the standards that we use for making judgments about the adequacy of the program and each of the human, decision, and technical matters involved. Is your program comprehensive, yet targeted? Is it creative in finding facts? Is it open to everyone who has something to contribute? Is it realistic and rational (does it meet standards of procedural rationality)? Is it integrative? Is it effective—that is, does it work in practice? Is it timely? Is it constructive, unbiased, and independent of special interests? Is it economical? Is it flexible? Is it responsible and honest, and does it have a reputation for honesty? In what ways does your program meet these standards, and where does it fall short? Where are the data to support your evaluation?

1) Describe a good program of gathering, processing, and sharing information. Is the current program meeting this standard? In what ways have research and the transfer and application of information been ideal? In what ways have they not? Is information being collected on all relevant components of the carnivore conservation program and from all affected people?
2) How open is the discussion about the meaning or relevance of information? By what standards are meaning and relevance judged? Which participants (official or unofficial) urge which courses of action, based on what information, for what purposes? Are people keeping common interests in the forefront, or are special interests trying to subvert the process of collecting, analyzing, and interpreting information?
3) Are the guidelines, policies, or plans that result from the preceding research and debate adequate to conserve and manage the species? Are they efficient, effective, and equitable? What is the basis for your judgment? What would be in an ideal policy, guideline, or plan for a conservation program? Does the current program approximate the ideal?
4) What would be the best ways to implement legislative policies such as the Endangered Species Act and other, more local management plans and guidelines, such as those being used in the Yellowstone region for wolves? What would be the features of organizations ideally suited to carry out such programs? In what ways have the agencies that implement programs and management activities performed well? In what ways have they not?
5) What would be the best ways to appraise or evaluate implementation of the program as well as the entire decision-making process that led up to implementation? In what ways has appraisal of the program been done well? In what ways has it not been done well?
6) Have policies such as the present approach to carnivore management led to success? By what standards do you define success? What factors should be considered in judging if the animals and habitats are well managed, if affected people have been treated

fairly, if institutions have been strengthened and trust in them increased? How should policies and related management be changed as needed? What should happen in terms of policy and management after the present policy ends to ensure future conservation and long-term coexistence? Why?

IV. What is your standpoint?
We all have different personalities, values, philosophies, education, experiences, and loyalties that give each of us a unique standpoint or viewing angle on the world and the program of interest. There is no such thing as a truly "neutral" or "objective" person or organization, although most of us aspire to be as bias-free as possible. How we see people and explain their behavior (including our own), how we go about solving problems, and how we find personal and professional meaning in our lives are all directly affected by our standpoints. Being aware of your own and others' standpoints is essential to good analysis and problem solving. Knowing the answers to these questions may tell you about unconscious biases that you or others have.

1) What roles do you and other people play in the conservation program? Are you a scientist, technician, manager, advocate, advisor, decision maker, scholar, facilitator, observer, analyst, or concerned citizen, or do you play another role?
2) What problem-solving tasks do you carry out when performing your roles? Do you help set goals, determine trends, analyze the conditions that underlie the trends, project trends into the future, or invent and evaluate alternatives?
3) What factors shape how you carry out your role and tasks—culture, personal interests, personality type, disciplinary training, organizational affiliation, and previous experience?
4) Which roles or problem-solving approaches are you attracted to in the conservation program? Which approaches or roles are you not interested in? Why?

Some Final Thoughts

Our shared interest is in finding more effective and sustainable ways for people and large carnivores to coexist. This task will likely continue to be problematic if past trends are a guide to the future. One conclusion seems obvious: if we—the extended community of people concerned about carnivore conservation—persist in our old perspectives and practices, unproductive conflict will remain with us. Improving programs and processes will require that people and institutions shift gears conceptually and practically.

As we see it, there are three possible outcomes for any program or conservation effort. In the win-lose situation, a "solution" is found when the most powerful side wins at the expense of the losers. This seems to be the way that large carnivore conservation typically unfolds. In the compromise situation, contenders are clear about what they stand to gain or lose, and they work out a deal that minimizes deprivations. Compromise is also part of most carnivore programs in Greater Yellowstone today. Integrative, win-win solutions are achieved when a new framework of cooperation is devised and adopted. Integrated solutions go well beyond winner versus loser or patchwork quilt compromises. They involve genuine innovation that redefines the context and offers participants the possibility of satisfying their underlying demands with no one losing out. New perspectives and practices come out of integrated solutions.

Contributors

Denise Casey is secretary/treasurer of the Northern Rockies Conservation Cooperative in Jackson, Wyoming. She has worked on wildlife studies in the U.S. and Australia and has written two books on human encounters with grizzly bears and wolves as well as several children's books on natural history subjects. She has also edited a number of books.

Tim W. Clark is a professor (adjunct) of wildlife ecology and policy in the School of Forestry and Environmental Studies and a fellow in the Institution for Social and Policy Studies at Yale University. He is also president of the board of directors of the Northern Rockies Conservation Cooperative in Jackson, Wyoming. He has worked in the Yellowstone system and in western Wyoming since the late 1960s.

Greg McLaughlin is currently a soil conservation technician with the Natural Resources Conservation Service in Ft. Collins, Colorado. He previously served as a project coordinator and principal grant writer for the Living Sustainability Laboratory, a demonstration site on alternative energy, green architecture, and conservation-based land use projects in western Colorado.

David Mattson is a research wildlife biologist with the U.S. Geological Survey at the Colorado Plateau Research Station in Flagstaff, Arizona. He has studied large carnivores for more than two decades, focusing on puma ecology and human-puma interactions in Arizona and the conservation and behavioral ecology of grizzly bears in the Yellowstone ecosystem.

Lyn Munno is a conservation planning associate for The Nature Conservancy in Vermont. She worked as a program manager for Earthwatch Institute, an international conservation organization, and was also an analyst for the U.S. Environmental Protection Agency.

Karen Murray is a program manager at the Grand Canyon Trust, a nonprofit conservation organization in Flagstaff, Arizona. She has held several different field biology positions for state and federal agencies, including studying bear-human interactions in Yosemite, Lake Clark, and Denali National Parks, and researching black bears in Virginia.

Steve Primm initiated and continues to direct the Gravelly Range Grizzly Project in southwestern Montana. He has worked since 1992 on large carnivore conservation projects, mainly focusing on grizzly conservation in the U.S. Rockies, ranging from participatory conservation to field research and proactive conflict resolution in the backcountry.

Murray Rutherford is an assistant professor in the School of Resource and Environmental Management at Simon Fraser University in British Columbia. In previous research he examined the evolution of ecosystem management policy in the U.S. Forest Service, with a focus on the Bridger-Teton National Forest in Wyoming.

Dylan Taylor has worked for Defenders of Wildlife in Washington, D.C., where he investigated oil and gas development on public lands and advocated for habitat protections in the West, and for the Northern Rockies Conservation Cooperative in Jackson, Wyoming, where he focused on the contextual elements—people, institutions, and animals—that shape wildlife management policy in Wyoming.

Jason Wilmot is the executive director of the Northern Rockies Conservation Cooperative in Jackson, Wyoming. Jason spent more than 10 years in the Glacier National Park Area, where he worked on a variety of wildlife monitoring and research projects. He developed a conservation and stewardship plan for an area of grizzly bear habitat with the Nature Conservancy of Montana and the Blackfeet Indian Land Trust, and was a backcountry ranger in Katmai National Park, Alaska.

Index